Renewable Resources Information Series
Coastal Management Publication No. 1

INSTITUTIONAL ARRANGEMENTS FOR MANAGING COASTAL RESOURCES AND ENVIRONMENTS

REVISED SECOND EDITION

JENS C. SORENSEN* AND SCOTT T. McCREARY**

*Department of Marine Affairs,
University of Rhode Island

**Department of Landscape Architecture
University of California at Berkeley

National Park Service
U.S. Department of the Interior
and
U.S. Agency for International Development

Second Printing
1990

The opinions, findings, conclusions, or recommendations expressed in this report are those of the authors and do not necessarily reflect the official view of the National Park Service or the Agency for International Development.

Sorensen, Jens C.

Institutional arrangements for managing coastal resources and environments. (Renewable Resources Information Series. Coastal Management Publication No. 1.)

National Park Service, U.S. Department of the Interior, Washington, D.C.

Cover title: Institutional Arrangements for Managing Coastal Resources and Environments.

Bibliography
   1)   Coastal zone management--developing countries.
   2)   Natural resources--developing countries.

     I.   McCreary, Scott T.
    II.   United States (National Park Service.)
   III.   Series.
   IV.   Title.
    V.   Title: Institutional Arrangements for Managing
              Coastal Resources and Environments

HT395.D44S671984; 333.91'7'091724; 84-17986   ISBN 0-931531-00-4

Available from:

Coastal Resources Center
Bay Campus
The University of Rhode Island
Narragansett, R.I. 02282

or

The International Affairs Office
U.S. National Park Service (023)
Washington, D.C. 20240

FOREWORD

Most countries recognize their coastal zones as distinct regions with resources that require special attention. Many have taken specific actions to conserve coastal resources and to manage coastal development. A few have created comprehensive nation-wide coastal zone management programs that are fully integrated with other resource conservation and economic sector programs. There is a current trend among the coastal countries to move toward more comprehensive and integrated coastal programs. To explore the results of this trend, the authors have reviewed the literature on institutional arrangements for coastal zone management in 75 countries, with concentrated attention on 25 of them. In so doing, they have produced the most detailed analysis of the subject yet prepared.

This book is one in a series of publications produced for the U.S. Agency for International Development (USAID) by the National Park Service (NPS) to guide the planning and management for sustainable coastal development and for the conservation of coastal resources. In addition to this book, the series includes a case book with eight case studies, a coastal protection guidebook and a program development guidebook.

This coastal series is part of a wider publication and training partnership between USAID and NPS under the "Natural Resources Expanded Information Base" project commenced in 1980 in response to world-wide critical need for improved approaches to integrated regional planning and project design. The project is producing publications on arid and semi-arid rangelands and humid tropic systems as well as on coastal zones. The publications and training components are dedicated to strengthening the technical and institutional capabilities of developing countries in natural resources and environmental protection and to providing other international development assistance donors with ready access to practical information.

The goal of integrated planning is to prepare a comprehensive plan in which the various development sectors have been assessed for their effects on the resources in given geographic areas (of which the coastal area is one of the most distinctive). In a world of rapid population growth and diminishing natural resources, countries that fail to plan their economic development strategy in concert with resource conservation and environmental management may not be able to sustain progress in health, food, housing, energy, and other critical national needs. Each developing country must have a realistic plan for accommodating its share of the 100 million people per year being added to the world's population. Such basic resources as fuel, water, fertile land, and fish stocks are already in short supply in many countries and their future prospects are in grave doubt.

While the presence of integrated planning and comprehensive management alone may not assure a sustained and ample yield from the coastal natural resources of any country, its absence will lead to their depletion. The opportunities for development based on excessive exploitation of coastal natural resources are rapidly fading. The future depends on development closely linked to resource conservation. In the coastal zone, the need for an enlightened approach is urgent.

Foreword

As noted by the authors, coastal zone management is a relatively new field that has its own special phraseology and concepts. The authors define coastal zone as "the interface or transition . . . that part of the land affected by its proximity to the sea and that part of the ocean affected by its proximity to the land . . . an area in which processes depending on the interaction between land and sea are most intense." They define coastal management as "any governmental program established for the purpose of utilizing or conserving a coastal resource or environment . . . and is intended to include all types of governmental intervention." Further, "the term implies that the governmental unit administering the program has distinguished a coastal area or zone as a geographic area apart -- yet between -- the ocean domain and the terrestrial or interior domain."

In producing the coastal publication series for USAID, we realize that we have, at best, provided a foothold for natural resource aspects of the new and rapidly expanding field of coastal zone management. Much important work lies ahead in many technical areas. We particularly recognize the need to provide specific natural resource working materials for regional planners and economic development planners. Also, there is a need for advice on protection of life and property against storms and other coastal natural hazards. Equally important is advice to planners on the role for designated protected areas-- reserves, parks, sanctuaries -- in tourism enhancement, fish stock management, and critical area and species conservation. We hope the present series will provide a springboard for studies on these important matters.

John Clark managed the coastal components of the NPS/AID projrct. We are especially grateful to William Feldman, Molly Kux, and William Roseborough, of the Office of Forestry, Environment, and Natural Resources of the Bureau of Science and Technology, for their continuing encouragement and patience.

Robert C. Milne
Chief, Office of International Affairs
National Park Service
Washington, D.C.

PREFACE

This book presents available strategies to strengthen the governance of coasts and the management of renewable natural resources in coastal zones of the developing nations. It represents our synthesis of literature from the developed and developing world along with the findings of our interviews with coastal resource managers.

Chapter 2 presents a series of technical terms and phrases to set the stage for the analysis that follows. Chapter 3 explains key differences and commonalities among coastal nations. The evolution of coastal management programs is explained in Chapter 4. Chapter 5 reviews coastal issues, and Chapter 6 identifies key actors in coastal resource management.

Chapter 7 presents eleven management strategies, and considers the advantages and disadvantages of each. Chapter 8 presents an array of alternate institutional arrangements and supplements. Program evaluation is discussed in Chapter 9 as the last step in the development of an integrated and comprehensive coastal resources management program. Chapter 10 offers two sets of recommendations. The first set is offered to guide the work of international assistance organizations, while the second set gives suggestions for the creation of national coastal resources management programs.

A key finding of this report is that **there is an array of management strategies and an array of institutional managements available** to help allocate coastal resources among competing and conflicting interests.

Our second finding is that **institutional arrangements and management strategies must be tailored to the needs of each individual coastal nation.** They should reflect the geographic scope of issues, and the existing institutions, political traditions, and technical capabilities of a nation.

A third key finding of this report is that **nations with one or more of four critical coastal-dependent sectors have a strong incentive to pursue integrated coastal management.** These sectors are: fisheries, tourism, mangrove forestry, or an economy vulnerable to coastal hazards.

We wrote this book for five audiences:

o **government officials** who administer coastal resources management programs, particularly those who now administer or may initiate integrated programs;

o **officials in international assistance organizations** who are concerned about the management of coastal resources;

o **staff and members of non-governmental organizations** that have a vested interest in the use of coastal resources and environments;

o **environmental policy consultants** who advise national

and international organizations on coastal resources
management;

o   **scientists and other academicians** who conduct applied
research on coastal resources and coastal environments.

This is the second edition of this book. The first edition was published in
1984. That edition underwent three cycles of review and comment on two
complete drafts. The first draft was prepared as a discussion paper for a
workshop convened in November, 1983, by the International Affairs Program of
the National Park Service, and attended by individuals experienced in
international environmental management. The comments generated by this
meeting were incorporated into a second review draft. The second draft was
selectively distributed nationally and internationally for review and comment by
individuals who had been engaged in international coastal resources
management. Comments on the second draft were incorporated into a third
draft, which became the first edition of this book.

Many people contributed to the first edition by providing detailed comments
on the review drafts. They include: John Clark, Random Dubois, Daniel Finn,
David Fluharty, Charles Getter, John Horberry, David Kinsey, Molly Kux, Crane
Miller, James Mitchell, Renee Robin, Christine Rossell, Harvey Shapiro, Samuel
Snedaker, Paul Templet, and Stella Vallejo. We are especially grateful to Niels
West for his assistance in determining the sovereignty status of coastal
nations. John Clark of the National Park Service provided valuable advice,
support, and encouragement throughout the project.

We owe a special debt of gratitude to Marc Hershman for his collaboration
on the first edition. He also made a detailed review of the first edition and
offered recommendations that helped us improve the organization and the
presentation of our material.

Six years have now elapsed since the first edition. The book was widely
distributed throughout the world under the auspices of the U.S. Agency for
International Development and the International Affairs Unit of the U.S.
National Park Service. This revised second printing reflects the comments
received from colleagues as well as reviews in professional publications.

Five years ago the United Nations Ocean Economics and Technology Branch
(now reorganized into the Office of Ocean Affairs and Law of the Sea)
contracted Jens Sorensen to prepare a working paper on the resolution of
coastal and marine use conflicts in the developing world. Jens Sorensen and
Scott McCreary worked together on the United Nations contract. The work on
the U.N. contract enabled us to further develop our concepts on governance
arrangements (U.N. Office of Ocean Affairs and Law of the Sea, forthcoming).
We have drawn on this U.N. work to revise and update these concepts in
Chapter 8. Our work in the field also made us aware of the need to add a
chapter in this edition on competing interests, the actors in allocation and use
of coastal resources and environments. Chapter 6 is the addition on competing
interests and actors or stakeholders.

Gretchen Lovas provided final editing and production assistance for this
revised edition.

In the intervening years since the first edition was published we have had the opportunity to test the frameworks we described in the first edition in the real world of coastal management programs. The tests have occurred in programs in the United States, Latin America, Australia and West Africa. The frameworks on issues, governance arrangements and management strategies were used to structure two publications on Latin America. A special issue (Vol. 15, #1) of the **Coastal Management Journal** on coastal management in Latin America included articles on the Galapagos Archipelago, Argentina, Brazil, Ecuador, and Mexico. A collateral report will be published by the Organization of American States and will also include chapters on Colombia, Costa Rica, Chile, Panama, Peru, and Venezuela.

In Australia, five coastal states collaborated to develop a national issues index modelled after the framework presented in Chapter 5 and Appendix B. The list is intended to serve as an important guide in structuring new coastal management programs.

The core descriptions of coastal management programs, together with the chapters on management strategies and institutional managements, helped structure the first workshop on integrated coastal resource management in West Africa. That workshop, convened November, 1987 in Mbour, Senegal, brought together representatives from 12 nations. The results are described in another U.S. NPS/USAID publication, **Prospects for Integrated Coastal Resources Management in West Africa** (Clark, McCreary, and Snedaker, 1988).

Our experience with coastal zone management programs during the six years between publication of the first edition and this edition have enabled us to draw two conclusions. First, the frameworks we present in this book are useful to **structure program design**. Second, the frameworks can be used to **structure comparative assessments** across programs to determine their relative strengths and weaknesses.

Jens Sorensen
Scott McCreary

TABLE OF CONTENTS

Table of Contents

LIST OF FIGURES AND TABLES

1. INTRODUCTION

1.1 Objectives

This book characterizes the management strategies and institutional arrangements available to coastal nations to conserve and develop their coastal resources and to resolve coastal use conflicts.

We define **institutional arrangement** as the composite of laws, customs, organizations and management strategies established by society to allocate scarce resources and competing values for a social purpose, such as to manage a nation's coastal resources and environments. Over time, every coastal nation has established its own institutional arrangement for managing coastal resources and environments.

Our initial intent was to examine how integrated management of coastal resources as currently practiced by a number of governments may apply to developing coastal nations. This initial framework was broadened to include an explanation of the available management strategies and institutional arrangements used to guide coastal resource use in all coastal nations. This expanded scope helped us achieve three purposes:

   o   present and comparatively assess the full range of
        national and subnational approaches to coastal
        resources management;

   o   guide the choice of coastal management strategies
        and institutional arrangements for the design and
        implementation of coastal resources management;

   o   propose a format for organizing information to
        facilitate information exchange.

Although the scope of analysis was expanded, our primary audience continues to be planners, administrators, scientists and senior officials interested in managing the renewable coastal resources in developing countries.

1.2 Work Program

We began our work on the first edition of this book with a review of the literature describing coastal resources management and environmental management in developing nations. Our literature review was supplemented by interviews with individuals who have had international experience in coastal resources management. No case studies were conducted for analysis nor were any visits made to coastal nations. In the six years since publication of the first edition, however, we have traveled extensively in both developed and developing coastal nations on research and consulting projects.

Three documents we examined were invaluable sources of information on coastal resources management programs. The documents are: United Nations

1

Ocean Economics and Technology Branch, **Coastal Area Management and Development**; James Mitchell, "Coastal Zone Management: a Comparative Analysis of National Programs"; and the report of USAID's five nation site visit to assess the potential for coastal resources management. These three documents are cited respectively as "UNOETB, 1982a"; "Mitchell, 1982"; and "Kinsey and Sondheimer, 1984."

2. CONCEPTS AND DEFINITIONS

Coastal zone management is a new field. Accordingly, there is no general agreement about the appropriate use or meaning of common phrases and terms. A number of terms are used interchangeably in the literature to describe the activity of managing a coastal region, area, use, or resource. These include **coastal management, coastal resources management, coastal area management, coastal area management and planning, coastal zone management, integrated coastal zone management, integrated coastal resources management and coastal zone resources management.** In general, these terms are not carefully defined or distinguished from one another, nor are the "resources" or "environments" that they manage well defined.

Given the global scope of coastal zone and exclusive economic zone management, it is essential to clarify terms at the outset. This section reviews some key concepts and definitions to establish a foundation for the analyses presented in later sections.

To acquaint the reader with terms in a logical order, the discussion begins with "coastal nations and subnational units" and "coastal management," then defines the natural areas and systems under consideration, and concludes with the specific management and planning terms.

Ten frequently used terms are defined in this section. They are:

- o   **coastal nations and subnational units;**

- o   **coastal management;**

- o   **coastal zone and coastal area;**

- o   **shorelands and coastal uplands;**

- o   **coastal resources, uses, and environments;**

- o   **coastal systems;**

- o   **coastal sectoral management or planning;**

- o   **integrated planning;**

- o   **coastal zone management program and integrated coastal resource management;**

- o   **ocean management.**

2.1 Coastal Nations and Subnational Units

We address four general categories of government authority which permit the establishment of an autonomous coastal management program. They are:

3

o   **independent nations** (referred to as sovereign states in the international law and political science literature);

o   **semi-sovereign nations, colonies, or dependencies;**

o   **subnational units** such as provinces, prefectures, or states empowered by the national constitution to undertake certain governmental functions, such as land use management;

o   **subnational regional authorities** established by legislative action or executive order.

The four distinctions were made to identify how many units of government there are in the world with the **potential** legal authority and resources to launch an integrated program for coastal resources management. According to the U.S. State Department's publication, **Status of the World's Nations**, the independence of Brunei increased the number of independent coastal (or ocean bordering) nations to 136 (U.S. Department of State, 1983). Since there are 30 landlocked nations, eighty-two percent of the world's independent nations border on the ocean or an ocean connected sea (Black Sea, Mediterranean Sea, Baltic Sea, Red Sea, and the Persian or Arabian Gulf). It is also noteworthy that 40 of the ocean or sea-bordering independent nations are small islands (e.g. Nauru, Barbados) or large islands (e.g. Papua New Guinea or Japan). This means that thirty percent of the world's independent nations are situated on large or small islands.

The State Department publication also enumerates "dependencies and areas of special sovereignty." Forty coastal semi-sovereign states have both sufficient resources and the population size to be self governing -- at least to the extent that the metropolitan nation could have granted them the statutory authority to establish their own coastal or ocean management program, examples of which include Bermuda, Hong Kong, and St. Kitts-Nevis. Of these 40 semi-sovereign coastal states, only five are not situated on large or small islands.

Several nations' constitutions delegate authority for specific government functions to the subnational level. Examples are the United States, Canada, Australia, and Malaysia. These nations have respectively 32, 8, 5, and 13 subnational units of government with authority to create an autonomous coastal management program. Coastal nations or subnational units also have the authority to establish by legislation regional entities to carry out coastal management programs. Examples are Australia's Port Phillip Authority and the San Francisco Bay Conservation and Development Commission (BCDC).

In sum, the number of government units in the world that have the legal authority to establish independent coastal management programs considerably exceeds the number of independent states. Taking into account the combination of 136 sovereign coastal nations, up to 40 semi-sovereign states, national subunits which are constitutionally autonomous, and regional authorities, on a world-wide basis the **potential** exists for creation of well over 200 and perhaps 250 distinct coastal management programs.

2.2 Coastal Management

Coastal management refers to any government program established for the purpose of utilizing or conserving a coastal resource or environment. It is the broadest of the terms used, and is intended to include all types of governmental intervention in a society. Use of the term implies that the governmental unit administering the program has distinguished a coastal area as a geographic unit apart, yet between, the ocean domain and the terrestrial or interior domain. The resources and/or environments being managed define the geographic extent of the coastal area (see Section 2.3 for a definition of "coastal area"). The coastal management program can consist of just one type of resource, such as coastal fisheries, or one type of environment, such as tidal wetlands. It is more common however, for a coastal management program to include several types of resources and environments.

2.3 Coastal Zone and Coastal Area

The image evoked by the term "coastal" varies considerably. To some it connotes fish and wildlife, to others beaches and dunes, and to still others broad reaches of land and water. Most agree that the term "coastal" conveys the notion of a land-ocean (or estuary) interface.

The land-ocean interface has two principal axes. One axis lies parallel to the shoreline (or longshore). The other axis runs perpendicular to the shore (or cross shore). For the longshore axis, relatively little controversy arises as to the definition, since it does not typically cross environmental system boundaries -- with the exception of watershed systems. By contrast, there is considerable discussion about the cross shore axis of the coastal zone. The cross shore axis profiles a coastal zone of transition between the ocean (or estuary) environment and the terrestrial or inland environment (see Figure 2.2).

The coastal zone is commonly referred to as the interface or transition space between two environmental domains, the land and the sea. It has been defined as that part of the land affected by its proximity to the sea and that part of the ocean affected by its proximity to the land (U.S. Commission on Marine Science, Engineering and Resources, 1969). It is an area in which processes depending on the interaction between land and sea are most intense. One lengthy definition combines demographic, functional, ecological, and geographical considerations:

> The coastal zone is the band of dry land and adjacent ocean space (water and submerged land) in which land ecology and use directly affect ocean space ecology, and vice versa. The coastal zone is a band of variable width which borders the continents, the inland seas, and the Great Lakes. Functionally, it is the broad interface between land and water where production, consumption, and exchange processes occur at high rates of intensity. Ecologically, it is an area of dynamic biogeochemical activity but with limited capacity for supporting various forms of human use. Geographically, the landward boundary of the coastal zone is necessarily vague. The oceans may affect climate far inland from the sea. Ocean salt penetrates estuaries to various extents, depending largely

5

upon geometry of the estuary and river flow, and the ocean tides may extend even farther upstream than the salt penetration. Pollutants added even to the freshwater part of a river ultimately reach the sea after passing through the estuary (Ketchum, 1972).

Invariably, "estuary" or "estuarine zone" is used in connection with or as part of the definition of the coastal zone. The term "estuarine zone" means:

> an environment system consisting of an estuary and those transitional areas which are consistently influenced or affected by water from an estuary such as, but not limited to, salt marshes, coastal and inter-tidal areas, bays, harbors, lagoons, inshore waters, and channels, and the term "estuary" means all or part of a navigable or interstate river or stream or other body of water having unimpaired natural connection with open sea and within which the sea water is measurably diluted with fresh water derived from land drainage (U.S. Department of Interior, 1970).

Given the environmental, resource, and governmental differences among coastal nations and subnational units, there is considerable variety in the selection of boundaries to delineate both the seaward and inland extent of the coastal zone. For example, the inland definitions of the coastal zone range from those that include entire watersheds, to those that comprise only the immediate strip of shoreline adjacent to the water. Ideally, a coastal nation or subnational unit should set the boundaries of the coastal zone as far inland and seaward as necessary to achieve the objectives of the management program. Since the problems and opportunities that motivate the creation of a coastal zone management program vary considerably from one unit of government to another, the selection of coastal zone boundaries would also be expected to exhibit considerable variation among coastal nations as well as among subnational units.

Small island nations or subnational units present a specific problem in setting the inland boundary of the coastal zone or area. An analysis of island ecosystems defines small islands as environmental units that do not have an "interior hinterland or central core area that is essentially distant from the sea" (Towle, 1985). The study concluded that approximately 10,000 square kilometers -- about the size of Jamaica -- is the breakpoint between small and large islands. In an island of less than one thousand square kilometers, there is no point that is more than a one hour drive from the sea, and one could argue that the entire island is a coastal zone. Coastal zone management on small islands is essentially synonymous with nation-wide or regional resource management.

Figure 2.1 displays the set of options for delineating the inland and ocean boundaries of the coastal zone. Figure 2.2 depicts these boundary options along a profile across the costal zone. The figure also presents examples of programs that use different sets of boundaries. The boundary options depicted in Figures 2.1 and 2.2 are now being used by coastal nations or subnational units to set the width of the coastal zone or the ocean management area (Section 2.10 discusses ocean management). At one extreme, the coastal zone could extend from the oceanward edge of the exclusive

6

Figure 2. 1: Options for Delineating the Ocean and Inland Boundaries of a Coastal Zone or an Ocean Management Area

Increasing Jurisdictional Area →

Inland Boundary Options ↓ \ Oceanward Boundary Options →	Mean low tide (MLT) or mean high tide (MHT)	Arbitrary oceanward distance(s) from a tidal mark	Boundary between provincial or state jurisdiction and national jurisdiction*	Ocean boundary of the territorial sea* (usually between 3 and 12 n.m. from CB) †	Ocean edge of the continental margin or shelf**	Ocean boundary of the exclusive economic zone (EEZ) ††
Arbitrary distance(s) from a tidal mark (such as 200 meters from low tide)	Costa Rica (MLT)	Sri Lanka Brazil Israel	California (from 1972 to 1976)	Spain	Great Barrier Reef Marine Park Authority	Sri Lanka, Netherlands, and Sweden ocean management program
Inland boundary of local government's jurisdiction***	Western Australia (MHT)		State of Washington (for planning)			
Inland limits of lands on which adverse impacts may be generated			• U.S. Coastal Zone Management Act • California (since 1976)			
Inland limit of climatic influence						

← Increasing Jurisdictional Area →

* In many cases the boundary between a coastal state (or province) and the national jurisdiction is the same as the territorial sea boundary line.
** In a number of places the continental margin extends oceanward beyond 200 nautical miles.
*** The inland boundary of local government's jurisdiction often extends further inland than the lands on which adverse impacts may be generated.
† The coastal baseline (CB) is a series of straight lines that interconnect coastal islands, headlands and promontories. It is used to map the oceanward boundary of the territorial sea and the exclusive economic zone.
†† The EEZ extends 200 n.m. or to the oceanward limit of the continental margin, whichever is greater.

Figure 2.2: Existing and Potential Boundaries of Coastal Zone Management Programs and Ocean Management Programs

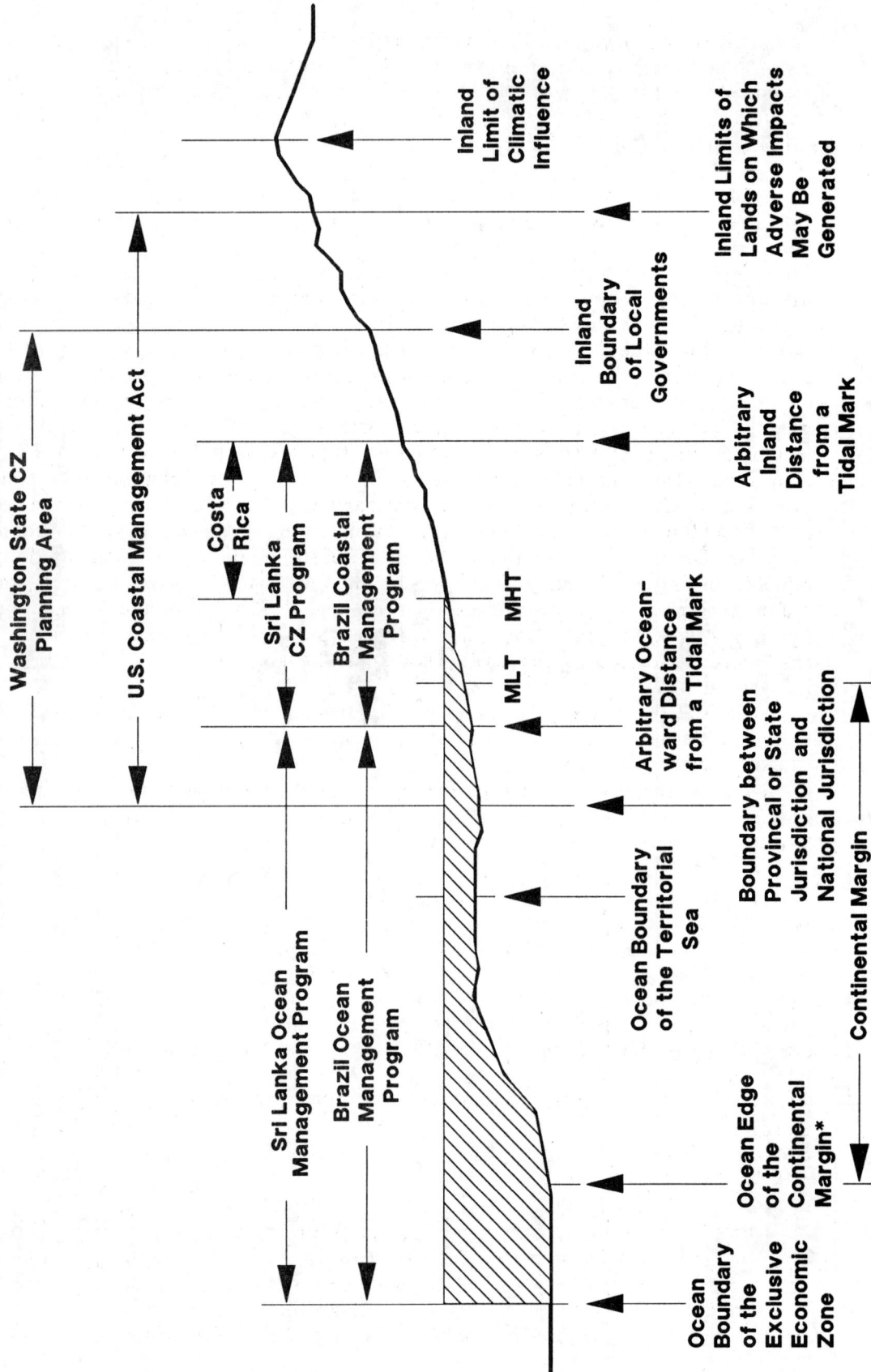

* In a number of places the Continental Margin extends oceanward beyond 200 nautical miles.

economic zone, usually 200 nautical miles (Section 2.10 discusses the exclusive economic zone), to the inland limit of climatic influence. This zone could measure at least 250 nautical miles wide. At the other extreme is a program -- such as Costa Rica's -- that extends from the mean low tide line to an inland distance of 200 meters.

Table 2.1 depicts the landward and oceanward boundaries for 13 programs. The most common inland boundary is an arbitrary distance from mean high tide and the most common oceanward boundary is the limit of state or provincial jurisdiction. The table also shows that a program can have at least two different inland boundaries, the planning zone and the regulation zone. The planning area should cover all lands on which development may generate impacts that significantly affect coastal resources or environments. In many cases the planning area extends a considerable distance inland (to the limits of an estuary watershed, for example). In the regulation zone the government has the power to issue or deny development permits. If there are two zones, the planning zone will always be larger than the regulation zone. The table also shows that the inland boundary of a program can change over time. In the case of California, the inland boundary line for regulation was set in 1972 as an interim measure of 1,000 yards from mean high tide. This boundary was kept for four years until the California Coastal Plan was implemented by new legislation in 1976. After four years the California Coastal Commission had much better information for establishing an inland boundary line. The new boundary is a variable line that has been set according to the various issues that were addressed by the California Coastal Plan.

It should be noted that most of the inland boundaries listed in Table 2.1 have one or more exceptions. For example, in Sri Lanka the inland boundary line extends two kilometers up coastal water bodies from their "natural opening" to the ocean. In California, the 1972 inland boundary line could not extend more than five miles from mean high tide in Los Angeles, Orange and San Diego counties.

One of the few maxims in the practice of coastal zone management is that the boundary lines should be determined by the issues which led to creation of the program. Generally this means a variable inland line. There are two reasons for the variable line. Different issues require different size management areas and the issues change as one moves along the length of the coast. For example, in northern and central California the boundary line swings inland five miles to control forestry and agriculture practices that can have a direct effect on the quality of coastal rivers and estuaries-- particularly rivers with salmon runs. By contrast, in urban areas the inland boundary is no more than 200 feet from mean high tide because the main concerns there are direct access to the shore and the visual quality of coastal architecture and public areas.

In this report the term "coastal area" refers to a geographic space that has not been defined as a zone. In other words, in coastal areas the inland and ocean boundaries to the zone have not been set or approximated. Use of the term merely indicates that there is a national or subnational recognition that a distinct transitional environment exists between the ocean and terrestrial domains.

Table 2.1: Inland and Ocean Boundaries of Coastal Zone Management Programs

	Inland Boundary	Ocean Boundary
Brazil	2 km from MHT	12 km from MHT
California -1972-1976 planning	highest elev. of nearest coastal mountain range	3 NM from the CB*
-1972-1976 regulation	1,000 yds from MHT	3 NM from the CB
-1977-present	variable line depending on issues	3 NM from the CB
Costa Rica	200 meters from MHT	MLT
China	10 km from MHT	15 meter isobath (or depth)
Ecuador	variable line depending on issues**	
Israel	1-2 km depending on resources and environment	500 meters from MLT
South Africa	1,000 meters from MHT	
South Australia	100 meters from MHT	3 NM from the CB
Queensland	400 meters from MHT	3 NM from the CB
Spain	500 meters from highest storm or tide line	12 NM (limit of territorial water)
Sri Lanka	300 meters from MHT	2 km from MLT
Washington State -planning	inland boundary of coastal counties	3 NM from the CB
-regulation	200 ft from MHT	

* For a definition of coastal baseline (CB), see Figure 2.1.
** Ecuador has six special management areas instead of a continuous zone. Inland width varies according to issues in each area. Limits will be set after the initial stages of planning.

MHT = mean high tide, MLT = mean low tide, NM = nautical mile

2.4 Shorelands and Coastal Uplands

Within many coastal zones, a further geographic subdivision is made for land immediately inland from the highest tideline, often called the shorelands zone. Shorelands are the terrestrial portion of the coastal zone where the inland connection to the shoreline and coastal waters is most apparent. In most cases, the seaward limit is mean high tide.

The inland extent of the shorelands varies. Several criteria are used to define the immediate and apparent connection to the coastline, depending on the public purpose the shoreland zone is intended to address. The following five criteria are a synthesis of standards drawn from U.S. coastal states' programs, Australian states' programs, and U.K. programs.

o   For **public access**, easy walking distance to the shore -- usually 300 to 500 meters -- is often the key determinant. A longshore dimension is often included, to provide for lateral access along the shore.

o   **Hazard avoidance** programs are often established in reference to bluffs, flood-prone areas, or areas with historic landslides.

o   **Protection of sensitive habitats**, such as wetlands, unstabilized dunes (those not stabilized by woody vegetation).

o   **Water quality protection** is achieved through setbacks for installation of septic tanks, and zones to keep natural vegetation along shores and banks -- both to control erosion and to retain the natural filtering capabilities of this vegetation. In this case the first tier of lots inland from the shore may be a logical shoreland boundary.

o   **Visual protection** of the coast is often accomplished with a shoreland zone defined in reference to the first public road paralleling the shore. Retention of natural vegetation along the shoreline is often a key element of such programs.

Exclusion zones and their applications are described in more detail in the chapter on management strategies (Section 7.6). Given the apparent land-ocean connection, shorelands are usually designated to provide government with more regulatory authority than in areas of the coastal zone that are further inland. Generally, a person anywhere in the shorelands area will be able to see, smell, and hear the ocean or a coastal water body such as an estuary or lagoon. As shown in Section 7.6, shoreland exclusion zones often use an average fixed inland limit. A number of programs have drawn the inland limit of the coastal zone at the shorelands/inland boundary. Usually this more conservative action is taken to gain acceptance for the coastal program.

Coastal uplands are defined as the area between the landward extent of the shorelands and the inland extent of land, the use of which could have a direct and significant impact on the quality of coast resources (see Figure 2.2). For small watersheds, coastal uplands extend to the inland boundary of the watershed. In cases where coastal mountain ridges are parallel and proximate to the coasts, coastal uplands extend to these ridgelines. Such a topographic configuration produces drainage patterns that affect the coast. It is also the inland barrier to marine climate penetration. Activities that often generate impacts in coastal uplands include road construction, new land clearance and development, agriculture, and logging (Dubois, Berry, and Ford, 1985).

2.5 Coastal Resources, Uses, and Environments

Within all **coastal areas or zones**, there are **coastal resources, coastal uses, and coastal environments**. A coastal resource is usually defined as a natural, often renewable commodity, the existence of which depends on the coast or the value of which is appreciably enhanced by its location within the coastal zone. Sometimes constructed features such as a scenic coastal village are included in the definition of coastal resource.

As we use the definition, the products of agriculture or forestry practiced near the shore are **not** coastal resources **unless** they achieve substantial production advantages from conditions associated with their coastal location. Similarly, land with a view of the ocean or within easy pedestrian access of the coastline can be considered a coastal resource since its value as property is usually enhanced by these attributes.

The types of natural or constructed features that fit the meaning of a coastal resource can be very broad. For example, the definition of coastal resources given by the **California Coastal Plan** (California Coastal Zone Conservation Commissions, 1975) divides coastal resources into several overlapping categories:

o   **Natural resources** - e.g., agricultural lands, coastal
      waters, beaches, clean air.

o   **Marine resources** - e.g., coastal waters, kelp beds,
      salt marshes, tidepools, islets and offshore rocks,
      anadromous fisheries.

o   **Coastal land resources** - e.g., watersheds, freshwater
      supplies, agricultural land, open space, bluffs, dunes,
      wildlife, natural habitat areas.

o   **Productive resources** - e.g., maricultural areas,
      gravel deposits, agricultural and timber lands,
      petroleum resources.

o   **Constructed resources** - coastal communities and
      neighborhoods with particular cultural, historical,
      architectural, or aesthetic qualities. These towns
      and neighborhoods are characterized by orientation
      to the water, usually a small-scale of development,

pedestrian use, diversity of development and activities, public attention to and use of facilities, distinct architectural character, historical significance, or ethnic or cultural characteristics sufficient to yield a sense of coastal identity and differentiation from nearby areas.

o **Historical and prehistorical resources** - e.g., recognized historical landmarks, outstanding architectural landmarks, Indian burial sites and shell mounds, plant and animal fossils.

o **Recreational and scenic resources** - e.g., beaches, coastal streams, marinas, SCUBA diving areas, scenic coastal roads, and other land and water areas with the potential for providing significant recreational use for the public.

o **Educational and scientific resources** - e.g., marine life refuges, rare and endangered species habitat, primitive areas, tidepools, wetlands.

The list includes many coastal environments such as watersheds, bluffs, dunes, islets, tidepools and salt marshes. Coastal environments are natural and constructed physical conditions that are either specific to the coastal zone (e.g., estuaries) or whose attributes are significantly determined by its coastal location (e.g., fishing villages).

The two terms are interconnected since the capacity of coastal resources to provide social utility is directly dependent on the conditions of the coastal environment. For purposes of policy-making, it is not important to draw a distinction between coastal resources and coastal environments. We use the term "coastal resources" in its broad sense to include coastal environments.

We note that developing nations will probably be most concerned with those coastal resources of direct economic or social value to its citizens. Visual and recreational resources of the coastal zone may be of lesser concern to developing nations unless coastal tourism is either an important economic sector or is an area for potential economic growth.

Coastal use refers to the utilization of coastal resources for economic, aesthetic, recreational, scientific or educational purposes. Use may be either consumptive or non-consumptive. For example, fishing is a consumptive use while bird watching is a non-consumptive use.

The distinction between coastal-dependent uses and non-coastal dependent uses is a cornerstone of most integrated coastal zone management programs. A coastal-dependent use requires an immediate coastal site to be able to function at all. Examples are fishing, mariculture, port facilities, offshore oil extraction, boat works, and marinas. An economic utility argument can be made to support the policy that coastal-dependent uses should not be preempted or precluded from shoreline or offshore locations by non-coastal dependent uses (such as residential development). Recently a study was done applying the concept of coastal dependency to case studies in five

Northeastern states (Marine Law Institute, 1988). The study provides a thorough analysis of this concept.

2.6 Coastal Systems

The aggregation of environmental and physical systems in a compact area is the distinctive characteristic of the coastal zone. In fact, the coastal zone has been defined by this aggregation of systems. At least nine major systems affect coastal management: (1) large-scale geomorphic or oceanographic units, (2) estuary watersheds, (3) estuary circulation systems, (4) ocean basins, (5) longshore circulation cells, (6) air basins, (7) populations of sport and commercial species, (8) viewsheds, and (9) public services. Of the nine systems, four are specific to the coastal zone: large-scale geomorphic units, estuary circulation systems, ocean basins, and longshore circulation cells. Five systems have hydrologic dynamics as the interconnecting mechanism. A recognition that these nine systems interconnect the coastal zone through impact networks must be a cornerstone of coastal zone management.

Some of the major issues these costal systems pose for coastal management are:

1. **Large-scale geomorphic or oceanographic units.**

   - the formation, growth, and decay of barrier islands, coral reefs, atolls;

   - sea level rise from global warming and/or subsidence or emergence of tectonic plates.

   - coastal ocean currents such as the Gulf Stream or the Humboldt Current.

2. **Estuary watersheds.**

   - ground water or surface water pollution, estuary water quality, and effects on flora and fauna;

   - ground or stream water flows, estuary and wetlands salinity, and effects on biota;

   - land use practices, run off, stream water flows, and stream or estuary flooding;

   - stream sediment loads, estuary sedimentation, and effects on biota;

   - stream sediment loads and deposition of beach materials on estuary or open coast shore (and then into the system of longshore circulation cells -- see #5).

3. **Estuary circulation systems.**

   - direct discharge of wastewater into coastal waters from all sources, estuary water quality, and effects on biota.

4. **Ocean basins.**

   - direct discharge of wastewater, oil, and solid waste from all sources, quality of ocean waters and sediments, and effects on biota;

   - estuary pollution, quality of ocean waters and sediments, and effects on flora and fauna.

5. **Longshore circulation cells, coastal erosion and deposition.**

   - control of coastal erosion and erosion-accretion dynamics within littoral circulation cells.

6. **Air basins.**

   - atmospheric emissions from all sources, ambient air quality, and effects on biota and human health.

7. **Populations of sport and commercial species.**

   - degradation of coastal streams and habitat of anadromous fish populations;

   - degradation of estuarine habitats and size of waterfowl, wildlife, and fish populations;

   - harvesting of commercial or sport species and maintenance of a sustained-yield population and food web.

8. **Viewsheds.**

   - development in areas visible from the first public road parallel to the coast, public recreation areas, or tourist facilities;

   - control of development in areas visible from major public use facilities;

   - design of guidelines for coastal development visible from recreation or tourism areas.

9. **Public service systems.**

   - land use within sewage services district and capacity of sewage system;

- land use within water services district and capacity of water supply system;

- land use within highway service area and highway congestion;

- land use and the ability to evacuate residents from storm hazard-prone areas before the advent of hurricanes, typhoons, or tsunamis.

Developing coastal nations probably will be concerned with management of those coastal systems which have direct and significant effects on the national economy or society. They may be less concerned with protecting coastal viewsheds unless coastal tourism is an important sector of the economy.

2.7 Coastal Sectoral Management or Planning

Coastal sectoral management or planning connotes the management of a single resource or use by a unit of government. For instance, a program focused on control of shoreline erosion in the coastal zone is a coastal sectoral management program. Sectoral planning is most often undertaken for ports, fisheries, tourism, oil and gas development, and wildlife (sectoral planning is discussed further in Chapters 7 and 8).

International assistance agencies use the term sectoral planning or management to describe a socio-economic development area. In the field of natural resources and environmental planning the most commonly conducted sectoral development programs are: agriculture, forestry, fisheries, energy, transportation, industrialization, urbanization, and public health and safety. This report divides a number of these eight sectoral development areas into more specialized coastal components. For example, transportation is divided into shipping, ports, and surface transportation. Table 8.1 lists typical sectoral divisions in coastal zone management.

2.8 Integrated Planning

Integrated planning is designed to interrelate and jointly guide the activities of two or more sectors in planning and development. In the context of coastal zone management, integrated planning usually implies the programmatic goal is to balance and optimize environmental protection, public use, and economic development. Often, integration also assumes coordination between data gathering and analysis, plan-making, planning, implementation, and construction. The two most common expressions of integrated planning are national economic planning and land use planning, also known in some countries as town and country planning (both are discussed in Chapter 7).

The recent focus on integrated planning and management reflects the growing awareness among developing nations that renewable natural resources are the foundation needed to build economic and social development programs. A 1979 report on environmental management in developing countries reinforces this view (USAID, 1979).

For the principally agricultural societies that predominate in developing countries, poverty and environmental degradation are in fact two manifestations of the same phenomenon: the unplanned, unmanaged impact of growing populations on a fragile natural resource base whose productivity is measurably diminishing in our own lifetime. If the material circumstances of the world's poorer people are ever to be improved over the long-term, ways will have to be found to husband the fragile natural resources upon which their well being depends.

Compared with inland environments, the coastal zone is more richly endowed with renewable natural resources. These include productive fisheries, soil and forests as well as the recreational values of coastal waters, beaches, and shorelands.

2.9 Integrated Coastal Zone Management

A five day workshop was convened in July 1989 to bring together individuals from all over the world who had been involved in coastal zone management programs. There were 28 participants from 13 different nations. The purpose of the meeting was to review progress in the last 20 years and consider the future of the practice. Two of the most animated points of discussion were the appropriate term to call the practice and a brief definition of the practice. There was general consensus that best name for the practice was integrated coastal zone management ("ICZM"). Alternative terms that have been used over the years were considered, including; coastal area management and planning, coastal zone management, integrated coastal resources management and coastal management. The term, coastal management, was rejected because it was considered to be too general. This opinion is reflected in the definition offered for this term in Section 2.2.

After considerable discussion the participants of the July workshop agreed on the following brief definition of integrated coastal zone management;

> a dynamic process in which a coordinated strategy is developed and implemented for the allocation of environmental, socio-cultural, and institutional resources to achieve the conservation and sustainable multiple use of the coastal zone (Coastal Area Management and Planning Network, 1989).

The workshop participants were in general agreement that an integrated coastal management program would have all of the following five attributes:

o   It is a process that continues over considerable time. ICZM is a dynamic program that usually will require continual updating and amendments. ICZM is not a one time project.

o   There is a governance arrangement to establish the policies for making allocation decisions and, if the program is implemented, a governance arrangement for making allocation decisions.

17

o   The governance arrangement uses one or more management strategies to rationalize and systematize the allocation decisions.

o   The management strategies selected are based on a systems perspective which recognizes the interconnections among coastal systems. The systems perspective usually requires that a multisectoral approach be used in the design and implementation of the management strategy.

o   It has a geographic boundary that defines a space which extends from the ocean environment across the transitional shore environments to some inland limit. Small islands may not have an inland limit.

The terms governance arrangement and management strategy are explained in the respective chapters devoted to these topics.

Many times it is difficult to determine whether a program is an integrated coastal zone management effort and or some other form of environmental planning and management. For example, should a national program for the planning and management of coastal parks and reserves be termed an ICZM effort or just a sectoral program? Generally, a systems perspective and multisectoral approach are the two key characteristics that serve to distinguish ICZM from other types of environmental planning and management programs.

Making the distinction between what are and what are not ICZM programs is critically important for conducting international or national comparative analyses. If lessons are to be learned from the experience of past and present ICZM programs, we must be able to define what we are looking for among the myriad of approaches to environmental planning and management. Questions will be continually asked on the number and location of ICZM efforts in the world. The answer, of course, depends on how the term, integrated coastal zone management, is defined.

In this book we use two terms as synonyms for integrated coastal zone management. These are coastal zone management and integrated coastal management.

2.10 Ocean Management

Ocean management involves national direction and control of "ocean space" including surface waters, the water column, the seabed, and the subseabed. The area covered by ocean management can extend from the inland limit of national jurisdiction (usually mean high or low tide) out to the ocean extent of its most seaward claim. The Convention on the Law of the Sea (CLOS) allows a coastal nation to claim an exclusive economic zone (EEZ). The coastal nation may have exclusive use of the EEZ in order to exploit the resources off its coasts -- particularly fisheries, oil, gas, and minerals.

Two different methods can be used to determine the oceanward boundary of the EEZ; a coastal nation may use either or both methods. One method is to measure 200 nautical miles oceanward from the coastal baseline (CB). The coastal baseline is delineated by a series of straight lines interconnecting the nation's headlands, promontories, and islands. The other method is to use the maximum extent of the continental margin (see Figures 2.2 and 3.1). The continental margin consists of the continental shelf, the continental slope, and the continental rise. In most places around the world the continental margin is less than 200 nautical miles from the coastal baseline. In this case the oceanward boundary of the EEZ is usually set at 200 nautical miles. In those relatively few places where the continental margin exceeds 200 nautical miles, the coastal nation may use the continental margin to set the oceanward boundary of the EEZ.

There is a major exception to these two methods. The exception is nations with a geographic boundary to one another of less than 400 nautical miles or nations that are interconnected by the continental shelf. In either of these two cases, the oceanward boundary is usually a midline set at points half the distance between the respective coastal baselines of the adjoining nations.

There is some semantic confusion in the literature regarding the distinction between coastal and oceanic areas. Oceanographers with a global perspective on the oceans generally consider all the area within the continental margin to be coastal waters. Coastal zone managers usually consider all areas beyond the oceanward extent of the territorial sea (usually from three to twelve nautical miles) to be oceanic. The simplest way to distinguish a coastal management program from an ocean management program is whether or not a terrestrial zone is included within the program's jurisdiction. The terrestrial area would be any lands inland of the mean high tide.

If this distinction is made, the multiple use zoning plans of the Great Barrier Reef Marine Park Authority (GBRMPA) would constitute an ocean management program. The Authority's jurisdiction extends from the low tide line to the edge of the continental shelf. The ocean boundary has been set by five longitude and latitude points that are approximately 200 meters in depth. The series of straight lines interconnecting the five points define a boundary that varies in width from 40 to 150 nautical miles. The boundaries enclose an area of 350,000 square kilometers, a jurisdiction larger in size than the United Kingdom or two-thirds the size of France. The marine park is not a national park in the conventional sense. The concept is that of a multiple use planning strategy which provides for the management of the entire jurisdictional area by zoning.

Ocean development or management plans are being considered, prepared or implemented by Brazil, Sri Lanka, Sweden, and the Netherlands. According to a recent publication only the Netherlands and the State of Hawaii -- in addition to GBRMPA -- have developed ocean management programs (Vallejo, 1989). Figure 2.1 shows the proposed or actual boundaries of several ocean management initiatives.

3. DIFFERENCES AND COMMONALITIES AMONG COASTAL NATIONS

Our literature review reveals many geographic, environmental, social, and economic similarities as well as differences among coastal nations. These similarities and differences affect the likelihood that an integrated coastal zone management program will be created in a given nation. A developing nation's coastal characteristics also suggest alternative governance arrangements and management strategies for coastal programs.

The six main characteristics we found to be useful in distinguishing a coastal nation's disposition to coastal resources management are presented in this section. These characteristics are:

o   geographic disparities (dimensions of coastlines and ocean claims);

o   coastal resource value (economic sectors linked to the coast, which influence the value nations attach to coastal resources);

o   concentration of development and population;

o   coastal orientation;

o   level of development; and

o   existing or potential government powers in the coastal zone.

3.1 Geographic Disparities

Coastal nations claim varying amounts of coastal and marine space within their jurisdiction. **Ocean Yearbook 3** presents a table with three coastal or marine geographic measures for 155 sovereign nations and semi-sovereign states (Borgese and Ginsburg, 1982). The measures are:

o   coastline in kilometers;

o   coastline/area ratio (expressed as coastline in kilometers divided by total land area);

o   hypothetical area encompassed by a boundary extending to the 200 nautical mile exclusive economic zone or to the limits imposed by the exclusive economic zone of neighboring coastal nations.

Table 3.1 lists the five nations with the highest and lowest values for each of the three geographic dimensions. Canada has by far the longest coastline, over 90,000 kilometers, Indonesia is second with nearly 55,000 kilometers, and

Table 3.1: Geographic Dimensions of Selected Coastal Nations

	Coastline (km)	Coastline/ area ratio	Area to 200 nm (in 100 kms)
GREATEST LENGTH OF COASTLINE			
Canada	90,908	.0091	1,370.0
Indonesia	54,716	.0287	1,577.0
U.S.S.R.	46,670	.0021	1,309.0
Greenland	44,087	.0203	147.3
Australia	25,760	.0033	1,854.0
SHORTEST LENGTH OF COASTLINE			
Monaco	4	2.0	NA
Gibraltar	12	2.0	NA
Nauru	24	1.4290	92.8
Tuvalu	24	.9231	211.5
Jordan	26	.0003	NA
HIGHEST COASTLINE / AREA RATIO			
Macao	40	2.5	NA
Maldives	644	2.2	279.7
Monaco	4	2.0	NA
Gibraltar	12	2.0	NA
Bermuda	103	1.9	NA
LOWEST COASTLINE / AREA RATIO			
Zaire	37	.0001	.3
Iraq	58	.0001	.2
Jordan	26	.0003	NA
Sudan	853	.0003	26.7
Algeria	1,183	.0005	NA
GREATEST EXCLUSIVE ECONOMIC ZONE			
U.S.A.	19,924	.0021	2,220.0
Australia	25,760	.0033	1,854.0
Indonesia	54,716	.0287	1,577.0
Canada	90,908	.0091	1,370.0
U.S.S.R.	46,670	.0021	1,309.0
SMALLEST EXCLUSIVE ECONOMIC ZONE			
Singapore	193	.3310	.1
Iraq	58	.0001	.2
Togo	56	.0010	.3
Zaire	37	.0001	.3
Belgium	63	.0021	.8

KEY: NA = No measure given   (Source: Borgese and Ginsburg, 1982)

the USSR has a coastline of over 46,000 kilometers. At the other extreme, Monaco has a 4 kilometer coast, Gibralter's is 12 kilometers, and the island nations of Nauru and Tuvalu each have 24 kilometers of coastline.

A more meaningful measure of the importance of the coast to a nation is the **ratio of coastline to total land area.** Small island nations or peninsula nations have the highest ratio. Nations with high ratios are likely to depend heavily on their coastal resources. Macao, the Maldives, Monaco and Gibralter all have a coastline/area ratio ranging from 2.0 to 2.5. Conversely, large nations with short coastlines have a ratio of several orders of magnitude less. For Zaire, the ratio is one kilometer of coastline to every 10,000 square kilometers of land area. Low values are also indicated for Iraq, Jordan, Sudan and Algeria. Nations with low coastline/area ratios are unlikely to depend heavily on coastal resources.

However, nations with comparatively short stretches of coastline usually place a high value on their ocean access because it is a very limited and scarce commodity. For example, Jordan, with a coastline of only 26 kilometers on the Gulf of Aqabah (Red Sea), and a coastline/area ratio of .0003, has developed a detailed management program for their small window on the world's oceans. Within a 26 kilometer stretch, Jordan has to accommodate its only ocean port, its navy, and its coastal tourism area. The argument can be made that nations with a large amount of coastline (and a high coastline to total land area ratio) and nations with very little coastline (and a low coastline to total land area ratio) place a higher value on their coastal zone than the vast majority of nations that comprise the middle of the range.

The U.S. claims the largest exclusive economic zone (ocean area to 200 nautical miles or to the limits imposed by other nation's EEZ) with a 2.2 million square kilometer claim. Australia has a 1.8 million square kilometer claim, and Indonesia claims 1.57 million square kilometers.

The easy-to-measure quantity indicators presented by Table 3.1 indicate the importance of the coast. A high coast/land area ratio, a large coastline or a large exclusive economic zone are good indicators of the **potential** existence and exploitation of coastal resources. The expected result is that the nation would accord high value to coastal and ocean resource management. However, area, length or shoreline/area ratios often do not reflect the value of the coastal zone to a nation. For example, the polar ice pack renders much of the ocean claim for Greenland, Canada, and the USSR unusable for fishing, oil exploitation, and shipping. Similarly, many tropical island nations have very large ocean claims of relatively low fishery value due to limited productivity of the waters. In the Near East and North Africa the majority of the coastline borders hot barren deserts, creating an environment that precludes most types of coastal development.

3.2 Coastal Resource Value

The best measure of the coastal zone's importance to a nation is the quality or **value of coastal resources** within the nation's jurisdiction. The value that nations attach to coastal resources is directly related to the economic contribution of these resources. We find this economic contribution is typically expressed with four measures, which are useful for most economic sectors. These measures are:

o   monetary value of coastal resource production;

o   export earnings of coastal resource production;

o   number of people directly or indirectly employed;

o   the cultural value of the coastal resource to serve
     dietary, religious or social needs.

To firmly establish these values for a series of developing countries, a review and analysis of statistical data would be required. Although such a review is outside the scope of this book, Appendix A lists the data that would be needed to derive such quantitative indicators of coastal value for several sectors of the economy. As a first step we recommend that the availability of this data should be assessed.

A team of economists (Pontecorvo et al., 1980) designed a conceptual and statistical model for calculating the aggregate value of ocean and coastal resources to the United States economy. The analysts extracted the information to make the calculations for this "Pontecorvo model" from the census and national income accounting system.

Based on their model, the aggregate value of the U.S. ocean sector for 1972, the most recent year . . . data was available, [was] $30.6 billion, comparable to agriculture at $35.4 billion . . . since the total U.S. GNP for 1972 was $1,171.1 billion, the ocean and coastal sector contributed 2.6% of the total (Towle, 1985).

The Ocean Studies Program at Dalhousie University adapted the Pontecorvo model to other coastal nations (Mitchell and Gold, 1982; Towle, 1985). The analysts calculated that ocean-related activity accounted for 33% of St. Lucia's GNP in 1978, 32% of Antigua-Barbuda's GNP in 1981, and 30% of Grenada's GNP in 1982. The Pontecorvo model as adapted by Dalhousie quantitatively compares and ranks the relative economic importance of ocean and coastal resources among nations (Towle, 1985).

The four dominant economic sectors in the coastal zone are **fisheries, tourism, ports, and oil and gas extraction. Hard mineral extraction** is a fifth sector where the value of coastal resources is apparent. **Agriculture and forestry** are two other sectors which may derive a production benefit from a coastal location. Certainly the forestry yields from mangroves represent an important economic sector in many countries. **Coastal hazards** do not represent a productive sector of the economy, but they can certainly exert a significant economic impact. Thus, a nation that has sustained economic loss due to flooding, wave damage, or shoreline erosion, may attach significant value to the proper management of coastal resources and processes. Control or reduction of environmental hazards is usually a component of the public health and safety planning.

Another indicator of the value of coastal resources to a nation is government commitment to develop a particular sector of the economy that is coastal-related. This might be evident in the creation of a special department for coastal management, a legislative or executive act, a plan for resource development, or the allocation of funds to implement sector development plans.

3.3 Concentration of Development and Population

A distinct measure of the importance of the coastal zone to a nation is the relative concentration of economic development and population. There is a positive correlation between an increase in coastal zone resource values and an increase in both the concentration of population and economic development. These three indicators can also be measured separately. Population growth and economic development can occur in the coastal zone without direct connection to coastal resources. Non-coastal related manufacturing and other basic industries commonly occur in the coastal zones, to take advantage of terrestrial resources, transportation and infrastructure networks, as well as easy-to-develop land.

In most coastal nations, national capitols and their surrounding metropolitan areas are within the zone that significantly affects coastal resources. Such capitols and their metropolitan areas usually originated as the nation's major port and owe their early development to port and related transportation functions. However, much of their subsequent growth is in the government and finance sections and is not port-related. The result is that the port economy often becomes a secondary sector of the metropolitan region it spawned.

3.4 Coastal Orientation

We have suggested several indicators to describe a nation's relationship to its coast: coastline length, coast-to-area ratio, size of ocean area claim, contribution of ocean and coastal resources to the national GNP, awareness of coastal hazards, institutional development for coastal-related sectors, and concentration of development and population. Some of these are better indicators than others, but each at least suggests national interest. We refer to the composite of these factors as "coastal orientation."

Clearly, there are degrees of coastal orientation. At one extreme are the small island states or nations, such as Bermuda, the Maldives, the Seychelles, and Niue. In these nations, virtually no part of their environment or economy is without coastal influence. At the other extreme are nations with a tiny fraction of coast/land area and little coastal development of coastal resources. Examples of this second category are Jordan, Zaire, Sudan, Algeria, Iran and Iraq. In these nations the coast is valuable because the multiple use demand is great and the supply of coast is small. A more precise accounting of coastal orientation could be derived from a data base organized around the indicators listed above. In the absence of this data, we propose a descriptive four-part typology which is outlined below.

1. **Small island nations.** All are coastal-oriented, given their large coast to land area, the strong dependence of their economies on coastal resources, concurrent lack of inland economic base, and the concentration of their population along the coast (this conclusion is supported by the Dalhousie study (Mitchell and Gold, 1982) and Towle (1985)).

2. **Large island nations.** All are coastal-oriented, but usually not to the same degree as small island nations. Large island nations almost always have capitols on the coast. These nations typically have a coastal or island resource base and a more dispersed population. Coastal hazards are likely to be a strong concern in countries such as Sri Lanka, Japan, Great Britain, New Zealand, Indonesia, Cuba, Japan, Madagascar, and the Philippines.

3. **Coastal-oriented continental nations.** These nations are often characterized by strong fishing, ports, tourism, or offshore oil and gas sectors. Most coastal-oriented continental nations have a strong concentration of population and economic development on the coast. Often the major metropolitan area or the capitol city is in the coastal zone. The United States, Nigeria, Senegal, Uruguay, Tanzania, Argentina, Libya, United Kingdom, Denmark, Sweden, and Ecuador are examples.

4. The fourth category is made up of **continental nations which are not coastal-oriented.** Economic development in these nations is directed at terrestrial resources in the interior and the size of the coastal populations is less than that of the interior. The USSR, Kenya, Germany and Argentina are examples. Many continental nations have major ports (e.g. Poland, Belgium, Germany) and some have distant water fleets (e.g. USSR, Poland, Romania) but otherwise have relatively little involvement with the coastal zone.

The best test for characterizing a nation's coastal orientation would be to apply the Pontecorvo-Dalhousie model to all nations. However, even if the necessary economic data were available for all coastal nations, which it is not, the time and effort needed to make the calculations would be prohibitively expensive.

It should be noted that the degree of coastal orientation can change quickly over time. International adoption of the 200 nautical mile exclusive economic zone will produce an increasing coastal orientation of the nations reaping large additional ocean and continental shelf area. Developing coastal nations have altered their orientation with new multiple year national economic plans significantly changing national investment in coastal and ocean resources development. An example is Uruguay's decision to develop its rich offshore fisheries. In 1974 the annual tonnage fished amounted to 12,000 tons. An ambitious multiple year fisheries development plan has increased the annual catch to 150,000 tons -- of which 85,000 tons are for export (Uruguay, Direccion Nacional de Relaciones Publicas, 1983).

3.5 Level of Development

International assistance organizations distinguish between developed and developing nations to set priorities for providing technical and financial aid. Several criteria have been used, such as GNP, annual per capita income, extent and quality of infrastructure, literacy rates, and institutional capacity. An assessment of environmental management in developing countries proposed a three part typology, dividing the world into developed nations, middle income developing countries, and lower income developing countries (International Institute for Environment and Development, 1981).

Developed nations have a per capita income per annum in excess of $1,000 and include 30% of the world's population (Riddell, 1981). In 1982 dollars, the range of middle income nations is between $200 and $1,000. Fifteen percent of the global population inhabit the middle income nations. Fifty-five percent of the world's population inhabit the low income nations where the per capita income per annum is below $200. The income levels have been found to reflect fundamental differences in (1) financial resources (or income levels); (2) economic and social infrastructure; and (3) available skills and knowledge (International Institute for Environment and Development, 1981). Some international government analysts have altered the hierarchy of income levels to create a discrete category for oil or mineral exporting nations which have surplus revenues but are otherwise not fully developed. Thus, the following four-part division is derived from the degree of development and surplus revenues:

o   developed or advanced income nations;

o   middle income developing nations;

o   developing nations with surplus oil or mineral
    revenues;

o   low income developing nations.

These levels of development may be important in guiding a nation's choice of strategies and institutional arrangements to establish coastal management programs (this is discussed further in Chapter 8 and Appendix E).

3.6 Existing or Potential Government Powers in the Coastal Zone

While the geographic scope and degree of control exercised by government authorities varies widely among nations, some general observations can be made on the relation between types of ocean or coastal areas and the degree of government control. Figure 3.1 illustrates the relative degree of national government influence for six different geographic areas: exclusive economic

Figure 3. 1: Extent of Government Control in the Management of Marine and Coastal Resources and Environments

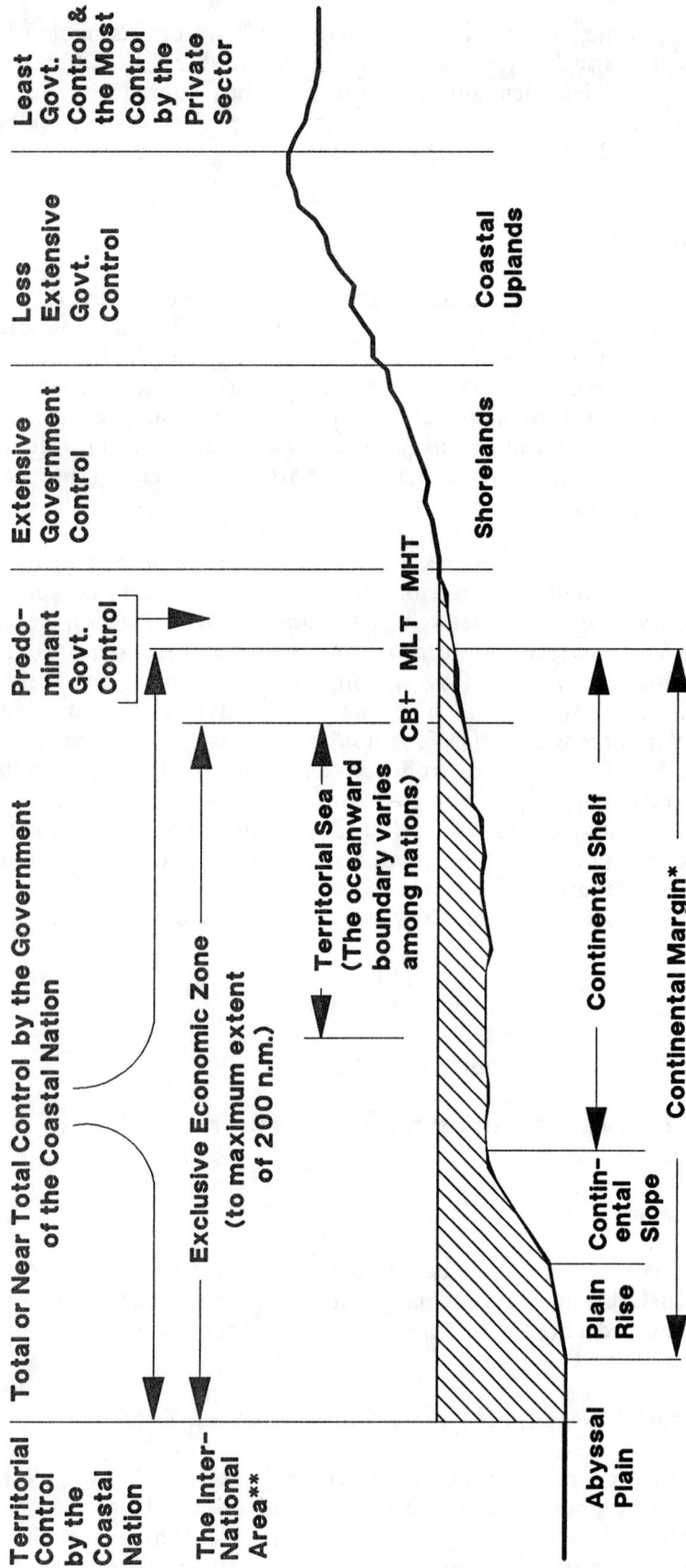

*In a number of places, the continental margin extends oceanward beyond 200 nautical miles. In these situations, the oceanward boundary of the national jurisdictional claim can be the outer edge of the continental margin.

**The international area is the seabed and ocean waters beyond either the continental margin or the exclusive economic zone (whichever is greater). Under the 1982 Law of the Sea Convention, the International Seabed Authority has some management authority, particularly with respect to marine mining.

†CB – The coastal baseline is a series of straight lines that interconnect coastal headlands and promontories. The CB is the reference point used to map the oceanward boundary of both the territorial sea and the exclusive economic zone.

MHT – Mean High Tide
MLT – Mean Low Tide

zone, territorial sea, continental margin, intertidal, shorelands, and coastal uplands. Within the territorial sea and exclusive economic zone, total or near-total government control is exercised. This control was reaffirmed by the Convention on the Law of the Sea which adopted the concept of a two hundred nautical mile exclusive economic zone (EEZ). The national government usually has most -- if not all -- of the powers to manage the exclusive economic zone. Some authority is frequently delegated to coastal subnational governments.

For the intertidal zone, the public trust is asserted, which in turn carries predominant government control. In many nations, the concept of the public trust is the heritage of common law. In Mexico, for example, the seashore up to the high tide line is "burdened with a right of commons quite similar to [American] . . . tideland trust" (Dyer, 1972). The Mexican concept of "property of common use" (bienes) is equivalent to American public trust lands and includes the seashore waters, fisheries, and riverbanks (Dyer, 1972).

The next area inland, the shorelands, are often subject to extensive government control. Exclusion zones are sometimes imposed in this band (e.g. Costa Rica) to prohibit private encroachment into wetlands, beaches, or to guarantee unrestricted public access to the shore. Prohibitions or strong regulations also may be imposed to protect coastal views or maintain water quality -- as discussed in Chapter 7, on management strategies.

For the coastal uplands, the tradition in most nations is to exercise less control than in the more shoreward areas. Exceptions are nations with strong programs for land use planning or town and country planning (e.g., Great Britain), or nations with a major commitment to economic development in a specific coastal region, such as France's development plan for the Aquitaine, or Mexico's tourism plan and development of Cancun and the southern tip of Baja California.

Finally, areas which have traditionally enjoyed no government control with respect to coastal resources are usually located inland of the coastal watershed boundary or beyond the most oceanward jurisdictional claim.

The complexity of government in terms of sectors, functional divisions, and number of levels usually increases in relation to the level of economic development. In middle income developing nations authority is likely to be distributed among several ministries, which may suggest formation of an interministerial council to draw together existing bureaucracies. The alternative is to create a new institution (see Chapter 8). Central government in the developing countries often has greater power to control land use and private property. Chapter 7 mentions two notable examples of a developing nation's public control over land: Nigeria's nationalization of all land not in productive use, and Costa Rica's shorelands restriction zone.

The characteristics of coastal nations presented in this section exert a strong influence over the choice of institutional arrangements and management strategies. This relationship is developed in more detail in Chapter 8 and Appendix E.

4. EVOLUTION OF INTEGRATED COASTAL ZONE MANAGEMENT: FROM CONCEPT TO PRACTICE

Our review of integrated coastal zone management efforts indicates that nations (and subnational units) follow a similar process in the evolution of their programs. These steps begin with an initial awareness stage and culminate in program implementation and evaluation. Figure 4.1 diagrams this general process.

4.1 Incipient Awareness (Stage 1)

Political recognition by a nation or subnational unit of the need for an integrated coastal management program usually requires obvious coastal resource damage or extensive destruction from coastal hazards. These events are compounded by the occurrence of intense conflicts among different coastal use activities (e.g. recreation vs. oil refineries and power plants) and their associated interest groups. In other words, a nation's or subnational unit's coastal resources and environments usually have to exceed some threshold of resource degradation, natural hazard destruction, or conflict before government will take action.

A catastrophic coastal event can catalyze public notice and government consciousness of the need for integrated coastal resources management. The Torrey Canyon oil spill in 1967 taught France and other nations that "institutional arrangements were . . . inadequate to deal with environmental disasters of such magnitude" (Harrison and Sewell, 1979). Similarly, the well-publicized oil spill in 1969 off Santa Barbara, California from the blowout of an offshore oil well did much to bolster the citizens' campaign to enact state-wide coastal zone management legislation (Adams, 1973).

Degradation, destruction, and multiple use conflicts are nearly always preconditions for consideration of integrated coastal resources management. Descriptions of the genesis of coastal awareness in ten nations confirm this observation. The ten nations are: the United States (Englander, Feldman, and Hershman, 1977), England (Waite, 1980; Steers, 1978), France (Harrison and Sewell, 1979), Greece (Camhis and Coccossis, 1982), the Australian states (Cullen, 1982), Sweden (Hildreth, 1975), Ecuador (Ecuador, Armada y Las Naci nes Unidas, 1983), Sri Lanka (Amarasinghe and Wickremeratne, 1983), the Philippines (Zamora, 1979), and Thailand (Adulavidhaya et al., 1982).

Figure 4.1 indicates that awareness of prospects for ICZM has been stimulated by the international travel of government, industry, and academic representatives to national as well as international conferences. Visits by foreign experts such as the Regional Seas survey teams, and international assistance missions, advisors or consultants may also stimulate such awareness. For example, the report of the United Nations mission to Sri Lanka in 1974 that recommended creation of a Department of Coast Conservation had a marked impact on national policy (Amarasinghe and Wickremeratne, 1983).

31

Figure 4. 1: Evolution of Coastal Management: From Concept to Practice

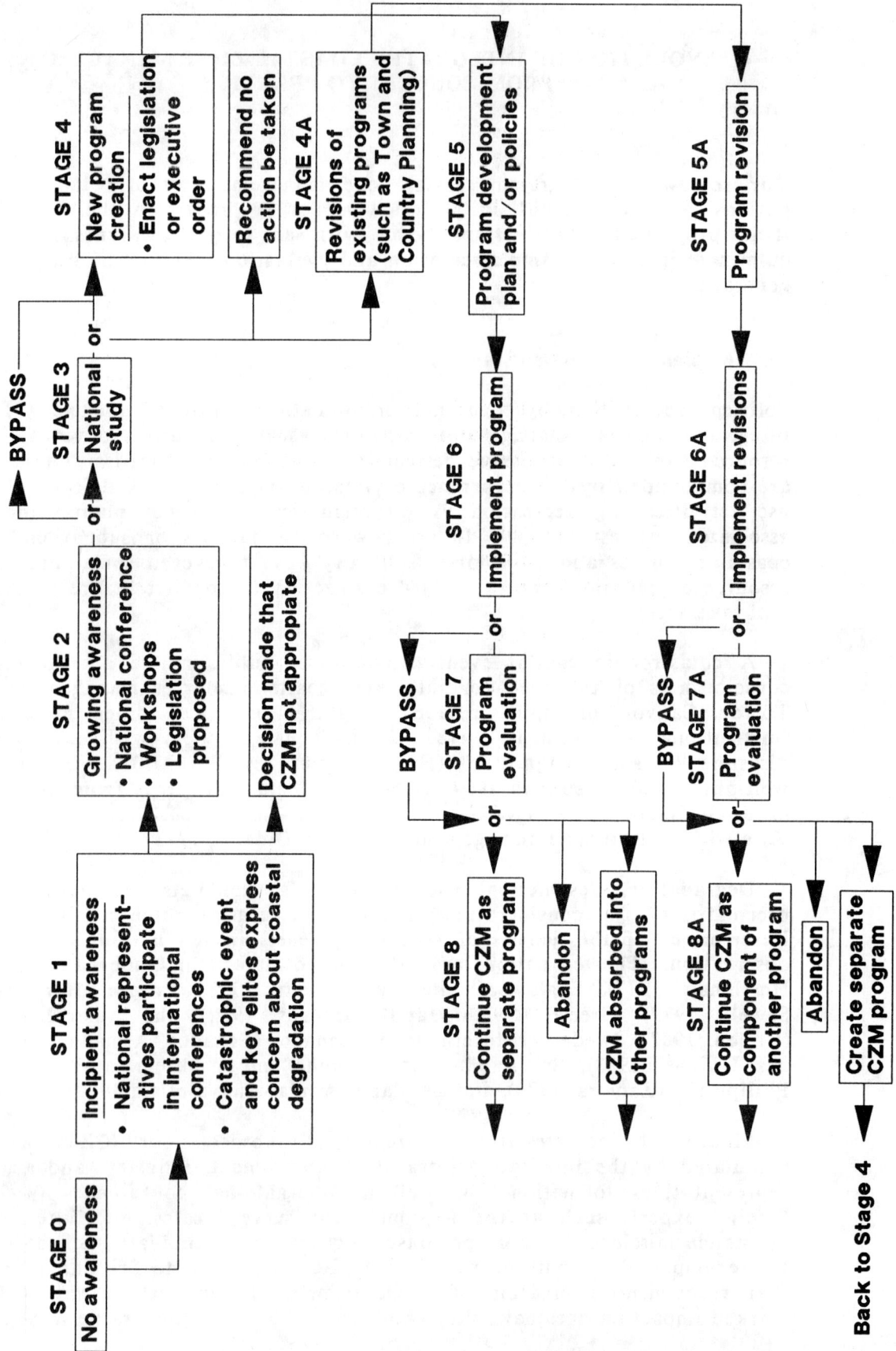

STAGE 0
No awareness

STAGE 1
Inciplent awareness
- National represent-atives participate in international conferences
- Catastrophic event and key elites express concern about coastal degradation

STAGE 2
Growing awareness
- National conference
- Workshops
- Legislation proposed

Decision made that CZM not approplate

STAGE 3
National or study

BYPASS

STAGE 4
New program creation
- Enact legislation or executive order

Recommend no action be taken

STAGE 4A
Revisions of existing programs (such as Town and Country Planning)

STAGE 5
Program development plan and/or policies

STAGE 6
Implement program

BYPASS

STAGE 7
Program evaluation

STAGE 8
Continue CZM as separate program

Abandon

CZM absorbed into other programs

STAGE 5A
Program revision

STAGE 6A
Implement revisions

BYPASS

STAGE 7A
Program evaluation

STAGE 8A
Continue CZM as component of another program

Abandon

Create separate CZM program

Back to Stage 4

4.2 Growing Awareness (Stage 2)

National conferences, workshops or hearings are usually the next step in program evolution. Such meetings or informal dialogues may be convened by government agencies, universities, industry associations or non-governmental organizations. Several nations have convened national conferences or workshops to consider the creation of integrated coastal zone management programs. These include the Philippines (Zamora, 1979), Australia (Cullen, 1982), Canada (Canadian Council of Resource and Environment Ministers, 1979), Ecuador (Ecuador, Armada y Las Naci nes Unidas, 1983), France (Harrison and Sewell, 1979), and Indonesia (Koesoebiono, Collier, and Burbridge, 1982).

4.3 National Study (Stage 3)

Conferences, workshops, or visits by international assistance missions often lead to the preparation of national studies. Such studies typically analyze coastal resources, institutional arrangements, and management options. The following eleven studies or proceedings illustrate this step:

o   U.S. Commission on Marine Science, Engineering and Resources (Stratton Commission), **Our Nation and the Sea**, 1969.

o   Great Britain, Countryside Commission, **The Planning of the Coastline: A Report of a Study of Coastal Preservation and Development in England and Wales**, 1970.

o   Sweden, Ministry of Physical Planning and Local Government, **National Physical Plan: Management of Land and Water**, 1971.

o   Ireland, Bord Failte Eireann and An Foras Forbartha, **National Coastline Study**, 1972.

o   France, Interministerial Committee for Regional Development and Planning, **The Picard Report**, 1972.

o   United Arab Emirates, **Coastal Development Planning Study**, 1976.

o   Ecuador, Armada y Las Naci nes Unidas, **Ordenacion y Desarrollo Integral de las Zonas Costeras**, 1983.

o   Philippines, National Environmental Protection Council, **Proceedings of a Planning Workshop for Coastal Zone Management**, 1978.

o   Israel, Ministry of the Interior Planning Section, **The National Outline Scheme for the Mediterranean Coast**, 1978.

    o   Canadian Council of Resource and Environment Ministers, **Proceedings of the Shore Management Symposium**, 1978.

    o   Australia, House of Representatives, **Report on Management of the Australian Coastal Zone**, 1980.

Of course, some national or subnational units create new coastal programs without conducting comprehensive studies or convening exploratory conferences. Greece, Indonesia, and Thailand have taken this route.

4.4 New Program Creation (Stage 4)

At least five of the national studies listed in the previous stage initiated or revised their coastal zone management programs. Findings and recommendations of the Stratton Commission report, **Our Nation and the Sea**, prompted the drafting of the legislation that ultimately evolved into the U.S. Coastal Zone Management Act (Zile, 1974).

The Countryside Commission Report lead to creation of the Heritage Coast Program for conservation of natural areas with scenic attraction and recreational opportunities for the public (Cullen, 1984). The literature on the history of coastal management in England does not indicate whether national land use policy makers at one point in time made a conscious decision not to create a new and separate program for the integrated management of the United Kingdom's coastal resources. In any event, the United Kingdom's arrangement for coastal governance consists of a number of revisions to existing acts and programs -- most notably the Town and Country Planning Act, the Local Governance Act, the National Parks Program, and the Natural Area Preserves Program (Waite, 1980; Steers, 1978; Cullen, 1984). We represent this revision of existing programs for coastal zone management as in Figure 4.1 as Stage 4A. The choice of program revision is distinct from the creation of a new, separate program or the decision not to embark on any form of integrated coastal zone management.

In Sweden, the National Physical Plan of 1971 lead to amendments to the Building Act and the Nature Conservancy Act (Hildreth, 1975). The revisions structured a new master (land use) planning process for all municipalities and directed initial efforts to coastal areas and those inland lakes where pressure for leisure home development was the greatest (Hildreth, 1975). The Swedish response is another example of Stage 4A, marginal revisions of existing programs.

The Picard Report of 1972 proposed that the French government take five measures to secure "sound" coastal management:

    o   creation of the Conservatoire du Litteral;

    o   the protection of sensitive perimeters;

    o   the development of coastal bases for leisure and nature;

o   the preparation of marine resource and sea water
      use plans;

o   the preparation of regional coastal plans (France,
      Ministry of the Environment, 1980)

The latest report in English indicates that all five of these measures have
been implemented. The most "successful" measure to date has been the land
acquisition and management program of the Coastal Conservatoire (France,
Ministry of the Environment, 1980).

It should be noted that ICZM may be initiated first at the subnational
regional scale before going nation- or state-wide. One example of this "scale
up" process is the initiation of the San Francisco Bay Conservation and
Development Commission seven years prior to enactment of the California
program, and another is the creation of the Port Philip Authority twelve years
in advance of Victoria's coastal program (Cullen, 1982).

All the existing national programs we have studied have followed the
process diagrammed in Figure 4.1, but two other avenues may emerge. Ocean
management programs spawned by the Convention on the Law of the Sea, such
as Brazil's ocean resources planning programs, as discussed in Section 2.10,
may spin off a coastal zone component as a separate program (Brazil, Comissao
Interministerial para os Recursos do Mar, 1980). The second avenue could be
evolution from a coastal and marine research coordination program. Colombia's
program for coordination of government and university marine research may
become an example of this avenue (Knecht, 1983).

4.5 Program Development, Implementation, and Evaluation (Stages 5 through 8)

We have reviewed solid information on coastal program development and
implementation for eight nations: Australia, Costa Rica, Ecuador, the
Philippines, Sri Lanka, Japan, the United Kingdom and the United States.
Most literature covers the U.S. Coastal Zone Management Act and the various
state programs (particularly California).   Costa Rica's coastal management
program was initiated by a new law in 1977. Now, twelve years later, the
program is making steady progress in preparing and implementing regulation
plans at the local government level (Sorensen, 1990).   Sri Lanka's coastal
program dates from the 1981 passage of a law requiring the preparation of a
national coastal program by 1986.   The plan is now being implemented,
primarily by regulating development within the coastal zone which extends
from 2 km seaward to 300 meters inland (Lowry and Wickremaratne, 1989).
Ecuador is beginning to implement a national ICZM program (Olsen, 1987;
Ecuador Ministerio de Energia y Minas, 1988).

In 1978 Israel produced the National Outline Scheme for the Coast. The
plan defined a coastal zone and proposed uses for each section of the coast.
Various components of the plan were being implemented in 1984 and other
aspects of the plan required additional development if the effort was to fully
achieve its objectives (Amir, 1984).

What became of the national studies and planning initiatives in Ireland and
the United Arab Emirates is unknown. Further analysis should be undertaken

to determine the fate and present status of all national or subnational ICZM efforts. The successes and failures of past initiatives would be informative both to countries considering ICZM and to countries engaged in ICZM programs. International assistance agencies considering integrated coastal zone management would also benefit from this information.

5. COASTAL ISSUES

We define issues here as the matters in dispute and the opportunities that motivate the creation and implementation of a coastal resources management program. An effort to define and understand the nature of coastal management issues is central to coastal resource management. This will become more apparent as we describe how issues drive the field of coastal management in the following five areas:

o   program design;

o   program evaluation;

o   information exchange;

o   setting international assistance priorities;

o   defining the field of coastal zone management.

Issues must be understood to ensure that the institutional arrangement fits the problems that the program is intended to solve. For example, if the issue is the impact of watershed practices on coastal resources, then the jurisdictional boundary should include the watershed area that generates the problems and the institutional arrangement should include the agencies with the appropriate watershed management authority. Different choices would be suggested, however, if the major concern is the management of the immediate shoreline area including issues such as coastal erosion, tourism development, and public access. In this case, a narrow jurisdictional zone would be appropriate. The organizational arrangement should consist primarily of agencies that exercise control over shoreline uses.

The issues that motivated a nation to design a program are likely to reappear as the criteria for program evaluation. (This is further discussed in Chapter 8.) The essential question in evaluating a program is to what extent it resolves the issues that motivated its creation. The full-scale design of a coastal zone management effort should contain an evaluation of program implementation. Moreover, international assistance agencies are increasingly requiring program evaluation to assess the success of their investment.

The international exchange of information typically involves technology development and application, such as dredging equipment or fisheries gear (Kildow, 1977), as well as the issues to which the technology is applied. Several international networks have already been established for information exchange on coastal management issues. For example, the International Geographical Union has formed a commission on the coastal environment to exchange information on coastal geomorphology. More recently, USAID, the U.S. National Park Service, and cooperating universities have created the Coastal Area Management and Planning Network (CAMPNET). There is relatively little international information exchange, however, on the institutional arrangements of coastal management programs. The variation in combinations of environmental, socio-economic, political and legal factors give a particular national character to each coastal nation's institutional arrangement. As a result, opportunities are limited for international transfer of information on a particular institutional arrangement.

International assistance agencies look for the transferability of experience and products to other nations. In general, the more often an issue arises, the greater the potential transferability. International assistance agencies would benefit from understanding the relative importance of issues in each nation. Clearly, issues that are globally common and consistently high in national priority warrant more attention from the international assistance community.

Coastal zone management lacks a disciplinary identity. Clear distinctions are lacking as to which issues are addressed by coastal zone management and which are not. Generally, if a problem or opportunity arises from the use of a coastal resource, it is a coastal zone management issue. This definition includes a broad spectrum of issues. The best approach to a universal definition may be to compile a list of the coastal nations' concerns for the management of their coastal resources.

Our literature review reveals a pattern among the issues. A few common themes demonstrate how the issues provide an international structure to the field of coastal management. Virtually every coastal nation with a major metropolitan area bordering an estuary has an estuarine pollution problem which is usually the result of municipal sewage and industrial toxins. Estuary pollution occurs in all coastal nations irrespective of the degree of development or environmental and socio-economic conditions. Nearly every coastal nation that actively harvests its coastal fishery stocks has an overfishing problem. In coastal nations with substantial mangrove acreage, environmental analyses usually report stress from watershed practices, pollution, filling, and overharvesting of timber for fuel. Similarly, the litany of institutional problems recurs in each discussion of a nations governance arrangement. Integrated coastal resources management appears almost invariably to be motivated by inadequate information, lack of intergovernmental coordination and inadequate professional resources.

5.1 Need for a Global Issues Index

The need for a global perspective and the recurrence of issues suggests two conclusions regarding the organization of the field of coastal management: (1) there is a need for an international indexing system; (2) such an index could be created rather simply. Some systems to categorize issues have been constructed for nations and ocean regions. Notable examples are:

    o   "Coastal Zone Problems: A Basis for Evaluation,"
           (Englander, Feldman, and Hershman, 1977);

    o   **Environmental Problems of the East African Region,**
           (UNEP, 1982c);

    o   **Marine and Coastal Area Development in the Wider
           Caribbean Area: Overview Study,** (UNOETB, 1980);

o   **Man, Land and Sea**, (Soysa, Chia, and Collier, 1982);

o   "Coastal Zone Management in Australia," (Cullen, 1982);

o   **Ordenacion y Desarrollo Integral de las Zonas Costeras**, (Ecuador, Armada y Las Naci nes Unidas, 1983).

However, our literature search did not reveal any attempts to construct a global index of issues.

Our design for a global issues index builds on the six previous efforts which have organized the broad array of issues into groupings. Generally, distinctions have been made between the following four types of issues:

o   **impacts of one coastal area activity** (e.g., tourism development or filling wetlands) **on others** (e.g., decreased commercial fishing yields);

o   **coastal hazards** or impacts of natural forces (e.g., shore erosion, river flooding, ocean born storms) on coastal use activities;

o   **development needs or sectoral planning** (e.g., fisheries development plan);

o   **organizational process problems**, such as an inadequate data base or lack of coordination.

Each of the issues is discussed below. Appendix B presents a preliminary global list of the issues for the first three groupings: impact issues, hazards, and sectoral planning.

5.2 Impact Issues

Impact issues are the most difficult issues to define. Many environmental and socio-economic causal relationships among coastal use activities form a web of interconnections and untangling these impact issues requires the determination of separate cause and effect chains (these are commonly discussed in the literature as impact networks or trees). In general, environmental and socio-economic impacts are the end result of a four-step process:

o   **coastal land or water use** (e.g., tourism development);

o   **specific activity** (e.g., filling of wetlands);

o   **change in environmental or socio-economic condition** (e.g., reduced estuary productivity);

o   **impact of social concern** (e.g., decreased fisheries yield).

For an issue to be perceived as a problem, the causal chain must evolve to the final step in the sequence -- an impact on a social value, such as a decreased fisheries yield. Appendix B assumes all chains culminate in impacts on critical social values.

We have clustered the list of impact issues in Appendix B into the following ten sets. The number in parentheses indicates the number of more specific issues contained in each set:

o   estuary, harbor and near shorewater quality impacts (14);

o   groundwater quality and quantity (2);

o   filling of wetlands (including mangroves) (5);

o   mangrove impacts (5);

o   coral reef and atoll impacts (9);

o   beach, dune, and delta impacts (5);

o   fishing effort (2);

o   access to the shoreline and subtidal area (2);

o   visual quality (2);

o   employment (2).

It is notable that the adverse impacts primarily relate to issues of water quality, pollution, and ecosystem types (e.g., wetlands, mangroves, coral reefs). There is some redundancy between two categories: filling of wetlands and mangrove impacts. Although mangroves are one type of wetland, the ubiquity and importance of mangrove systems to most developing nations merits a separate grouping.

Of the 48 separate impact issues listed in Appendix B, 27 of them concern effects on fisheries yield and 17 of them concern effects on tourism and recreation attraction (there is some double counting, and we combined similar impact chains). **Fisheries conservation and the maintenance of tourism or recreation quality clearly emerge as the two main arguments for integrated coastal resources management.** These two coastal uses are affected by almost all of the other use activities listed. The economic importance of fisheries and tourism will strongly influence the extent to which developing nations will want to initiate coastal resources management programs. Mangrove forestry operated on a sustainable yield basis appears to be of secondary importance, but is a significant coastal-dependent sector in several nations. The greater the value of coastal fisheries, coastal tourism, and mangrove forests to the national economy and coastal populations, the greater the nation's interest in coastal zone management.

The list of impact issues clearly illustrates the zonal nature of the coast. Nineteen of the issue impacts occasionally or always originate in coastal watersheds -- often far inland from the shoreline (DuBois, Berry, and Ford, 1985). On the ocean side, ten of the issue impacts can originate offshore and move landward to adversely affect coastline or estuary environments (Hayes, 1985; DuBois and Towle, 1985). The many watershed-coast-ocean connections clearly demonstrate that the coastal zone is where use activities which affect renewable resources must be coordinated.

5.3 Hazard Issues

Hazard issues constitute a relatively clear set of concerns. We found that five types of hazards were distinguished by coastal nations:

o   shoreline erosion (Hayes, 1985);

o   coastal river flooding;

o   ocean born storms;

o   tsunamis;

o   migrating dunes (DuBois and Towle, 1985).

All five hazards are naturally occurring phenomena. However, coastal erosion, river flooding, and dune migration can be caused in some situations solely by use activities, such as residential development. More commonly, these natural phenomena are exacerbated by the additional effects of human use activities.

Our literature review shows that coastal hazards are another major economic stimulus which initiate coastal resources management programs. This leads us to the conclusion that prospects for development of a coastal zone management program in lower income developing countries are strongest where fisheries, tourism, or coastal hazards devastation are important concerns, or where there is an infusion of international assistance funds and expertise.

5.4 Developmental Needs

Development needs are expressions of sectoral planning interest in response to one or more problems or opportunities identified by the coastal nation. However, coastal nations want information on these topics, and therefore development needs should be included in an issue-based information system. Eleven types of development needs emerged from the literature survey:

o   fisheries;

o   natural area protection systems;

o   water supply;

o   recreation development;

o   tourism development;

o   port development;

o   energy development;

o   oil or toxic spill contingency planning (as a
     component of water pollution control plans);

o   industrial siting;

o   agricultural development;

o   mariculture development.

Most coastal nations will prepare sectoral plans for fisheries, water supply, natural areas, port development, industrial siting, and agriculture. A meaningful difference among coastal nations will be the priority each assigns to the respective sectors. For example, is fisheries development planning higher on a nation's priority list than port development? The utility of the list would be improved if a number of broad sectoral categories were subdivided into more specific topics. For example, port development should be subdivided into the primary types of port facilities needed (e.g., oil, bulk container, general purposes, fishery, recreation marinas).

5.5 Organizational Process Problems

Analysts of program evaluation commonly distinguish between organizational process problems and outcome problems. We refer to outcome problems in coastal zone management as impact issues and hazards (Appendix B). Organizational process problems are procedures (or characteristics) that inhibit an organization from attaining its goals and objectives. A number of organizational process problems are also discussed in Chapter 9, Program Implementation and Evaluation.

An analysis of issues that motivated passage of the U.S. Coastal Zone Management Program (Englander, Feldman, and Hershman, 1977) identified five critical organizational problems:

o   **lack of coordination among public agencies;**

o   **insufficient planning and regulatory authority;**

o   **insufficient data base and lack of information for
     decision making;**

o   **little understanding or knowledge about coastal
     ecosystems;**

o   **resource decisions made primarily on the basis of
     economic considerations to the exclusion of
     ecological considerations;**

Eight other issues turned up as secondary in the analysis:

o   lack of clearly stated goals;

o   lack of state and local government funds to manage
     the coastal zone adequately;

o   primitive analytical tools and predictive
     methodologies;

o   dominance of short-term management over
     long-range planning;

o   complex, conflicting, and confusing laws;

o   little awareness of or concern with coastal
     problems;

o   lack of properly trained and educated management
     personnel;

o   limited public participation in decision making.

Few descriptions of coastal issues in developing nations discuss organizational process problems. Those that do usually identify problems such as lack of coordination, an insufficient data base, lack of personnel, lack of clearly stated goals, and outmoded laws. It appears that the same types of organizational process problems will occur irrespective of the nation's level of development. For example, lack of adequate governmental coordination and inadequate information for decisions are two problems inherent in almost all comprehensive policy-making.

We can expect that nations will vary in the relative importance of each institutional problem. Developing nations, for instance, have stressed the lack of properly trained and educated personnel and complex, conflicting, and confusing laws as obstacles to program development. However, these two concerns were given a relatively low priority in the survey of U.S. problems.

5.6 National Listings

Based on our literature search, we list in Appendix B nations that have expressed concern in published reports about each respective issue. This provides a starting point for a complete list of nations for each issue. Individuals familiar with a nation's coastal zone could no doubt augment the list with other issues that have arisen in that nation (in Appendix B we explain five problems that contribute to gaps on our list).

This national list could be improved by the addition of priority rankings. Englander, Feldman and Hershman's (1977) article on U.S. coastal management problems is one of few studies that indicates the relative importance of each issue.

A cross-section of nations would be sufficient to ensure that the issues confronting coastal nations are identified. The 30 coastal nations in Appendix B with complete descriptions constituts such a cross-section of important variables. The sample broadly represents variations in level of development, global climatic zones, and continental locations.

5.7 Surveying National Issues

We envision that compiling a global index of coastal management issues would be a two-step process. The first step is to construct the initial list of issues based on a review of national descriptions (Appendix B represents the product of this first step). The second step is to complete a global survey of all coastal nations both to further refine the issues list and to complete the national lists. This task was beyond the scope of this study, but the process for conducting the global survey is outlined in Appendix B.

Certainly coastal nations should be encouraged to conduct their own survey and ranking of coastal issues. The Philippines, Indonesia, Australia, New Zealand and Ecuador have defined national concerns by convening national conferences or workshops and appointing task forces. Most coastal nations cannot be expected to go to this level of effort. It is more likely that one agency -- such as the national planning office or an environmental policy council -- will compose a ranked list of coastal issues.

We do not expect that review of additional national descriptions or national surveys will add a significant number of new impact issues to the list. The 30 nations used as the basis for constructing the list in Appendix B are a representative sample of the world's coastal nations.

Since we wanted to determine the consensus of developing nations, an issue impact had to be identified by a developing nation to be included in our list. However, further iterations of the issues index should also include issues that are of concern only to developed nations. At a minimum, the list would help document one of the differences between these two groupings of nations.

6. MAJOR ACTORS IN COASTAL MANAGEMENT

If integrated coastal management efforts are to allocate scarce resources among competing interests, then coastal managers must identify and work with representatives of key interest groups. Table 6.1 suggests the array of actors that may have an interest in resolution of many coastal conflicts.

As Table 6.1 indicates, stakeholders in coastal issues may be active at the local, regional, national, or transnational level. They may be well organized or poorly organized. Among the well organized stakeholders are government agencies, parastatal corporations, and private industry, scientific, and conservation organizations. Less organized groups of actors may include land owners, ethnic groups, and social classes. Of course, the latter groups may become well organized in the context of political parties or other formal organizations.

6.1 Well Organized Actors

6.1.1 Elected officials.

Elected officials may be important players in coastal resources management. National political figures may champion a particular development project or conservation initiative. For example, in Ecuador, President Febres Cordero appointed the Comision de Alto Nivel para el Plan Maestro Galapagos (High Level Commission for the Galapagos Islands Master Plan). Local and regional officials may also be important actors in coastal management because of the power they can exercise in the allocation of coastal resources or land use.

6.1.2 Political parties.

Although coastal resource policy is not often a major priority of a national political party, parties may be the major sounding board for groups of affected interests.

6.1.3 National and subnational agencies with sectoral responsibilities.

One major group of coastal actors include ministries, subministries, and other bureaus with sectoral responsibilities. As shown in Table 8.1, a nation might have over twenty important sectoral interests in the coastal zone. The coastal zone-specific sectors include navy and national defense, port and harbor development, shipping, fisheries, mariculture, tourism, research, and erosion control. Many other sectoral activities are not restricted to the coastal zone, but may depend on coastal resources in part, or may directly affect the coast. Examples include agriculture, forestry, fish and wildlife management, parks, pollution control, water supply, flood control and energy generation.

Sectoral interests are often represented by ministries or branches of the executive branch, as well as by regional subdivisions and functional divisions.

45

Table 6.1: Potential Actors in Coastal Resources Management in Developing Nations

| | Subnational | | | Trans- |
	Local	Region	National	National
WELL ORGANIZED ACTORS				
Elected Officials	O	O	O	NA
Political Parties	O	O	O	NA
Government Agencies	O	O	O	O
Parastatal Corporations	O	O	O	NA
Private Industries	O	O	O	e.g., MNCs
Industry or Labor Orgs.	O	O	O	NA
Lending and Aid Institutions			O	e.g., UN, USAID, IBRD
Scientific Community	Local Polytechnic	State University		e.g., IGU
Conservation Organizations			e.g., NATMANCOMs	e.g., IUCN Greenpeace
LESS ORGANIZED ACTORS				
Subsistence and Artisanal Resource Users	O	-----Generally not organized------		
Coastal Property Owners	O	-----Generally not organized------		
Ethnic Groups	---------May become organized-----------			
Social Classes	---------May become organized-----------			

KEY:
O: Grouping in Common
NA: Not Applicable
IBRD: International Bank for Reconstruction and Development
IGU: International Geophysical Union
IUCN: International Union for the Conservation of Nature and Natural Resources
MNC: Multinational Corporations
USAID: U.S. Agency for International Development
NATMANCOMs: National Mangrove Committees

Each sectoral division of government may have several functional divisions (Figure 8.1). Examples of functional divisions are the power to levy charges, formulate policy, construct new projects, and disseminate information. Even within the same sectors (e.g. fisheries, or tourism), the staff charged with implementing each specific function will have a specific agenda.

6.1.4 State owned enterprises and parastatal corporations.

In many nations, some part of a coastal-dependent sector may be nationalized. For example, in Indonesia, the state owned oil company, Pertamina, is important in the petroleum extraction sector. In the Republic of the Seychelles' tourism company, the Compagnie Seychelloise pour las Promotion Hoteliere (Cosproh), and the Seychelles National Fishing Company (Snafic) are important in their respective sectors.

6.1.5 Private industry.

The private industrial users of coastal resources parallel the government sectors concerned with coastal resource use. Examples are fishing, tourism, ports, timber harvesting, oil extraction, lumber extraction, and salt harvesting. Malaysia's Matang forest supports mangrove harvesting on a 30 year rotation (Saenger, Hegerl, and Davie, 1983). Ecuador's shrimp mariculture industry earns export revenues of 225 million U.S. dollars per year (Snedaker et al., 1986).

6.1.6 Multinational corporations.

Certainly large multinational corporations (MNCs) are important private actors in coastal resource allocation. In fact, a single multinational oil company or mining consortium may have greater resources and greater political clout than a national economic sector.

Bargaining between MNCs and developing nations has been taken up by a number of political scientists. A variety of conclusions has been reached, ranging from the suggestion that major oil companies completely dominate the notion of the best interests of nation, to the idea that nations and subnational units can gain strength by establishing linkages with MNCs.

Some events suggest that a range of outcomes is plausible, depending on the strength and economic diversity of the nation, and the resolve of its political leaders. For instance, while MNCs exert strong influence in ad hoc ocean policy for Indonesia and Malaysia, they have made concessions to state-owned oil companies and local fishing interests (Klapp, 1984). In Aruba (Netherlands Antilles), Exxon has unilaterally announced plans to close the refinery that provides half the island's income. With its economic base shaken, Aruba has reconsidered its goal to attain autonomy and full independence (**New York Times**, 1985). Clearly, Exxon would be a dominant actor in any new scheme to integrate development planning across coastal sectors for Aruba.

Smaller transnational entrepreneurs also have interests in allocation of coastal resources outside their own countries. Japanese interests have funded prawn trawlers in Malaysia (Klapp, 1984), while Ecuadorian investors are beginning to acquire interests in mangrove habitats in the Dominican Republic for rearing ponds. These private enterprises would have a stake in fisheries sector development or the prospective designation of exclusion zones around mangroves.

6.1.7 Assistance institutions.

Agencies of the United Nations and international assistance organizations are becoming important players in coastal resource management in developing nations. The United Nations, through its Secretariat and individual agencies (UNEP, UNESCO) provides guidance and assistance on a variety of coastal issues. The UNOETB (1982a) published a major book on coastal management. The U.S. Agency for International Development has funded a series of investigations on coastal resources management (Sorensen, McCreary, and Hershman, 1984; Snedaker and Getter, 1985; Clark, 1985), and sponsored at least a six year effort to assist coastal management programs in Ecuador, Sri Lanka, and Thailand (University of Rhode Island and USAID, 1987).

6.1.8 Scientific community.

Scientific research is an important activity in the coastal zone of many developing countries. In some nations, such as the Cape Verde Islands, most researchers hail from foreign universities or institutions (McCreary, 1985). Also, scientific groups often have their own agendas for coastal management. In the Seychelles, the scientific community played a major role in the redesignation of the island of Aldabra from a proposed airstrip to a research preserve (Stoddart and Ferrari, 1983), and in the Galapagos Islands, the Darwin Research Station is an important player in coastal resources management because of the expertise of its scientists and the information base it has amassed over the years (Broadus et al., 1984; Broadus, 1985).

6.1.9 Conservation organizations.

Conservation organizations concerned with coastal resources can be major actors at the local, regional, national, and international level. Grassroots organizations such as Greenpeace, independent institutes such as the International Institute for Environment and Development (IIED), and hybrid organizations such as the International Union for the Conservation of Nature and Natural Resources (IUCN) are all important actors. Many of these groups, like their scientifically-oriented counterparts, are often referred to as "non government organizations" (NGOs), with the implication that they are neither bureaucratic agencies nor private for-profit firms.

Initially, conservation NGOs had to struggle for participation. Only one NGO attended the meeting of the International Whaling Commission (IWC) in 1964; in 1980 there were 40 present. But NGOs have lobbied effectively at the IWC and international fora such as the Convention of International Trade on Endangered Species (CITIES), and the London Dumping Convention. In many cases, conservationist NGOs were able to make their views known to national elites or members of the aristocracy who favored conservation (Stoddart and Ferrari, 1983). Now, different types of NGOs are forming more intricate coalitions and networks with consumer groups and other local interests to increase their influence and bargaining power (Barnes, 1984).

At least four types of conservation NGOs are now active in developing nations' coastal management programs:

o   national level organizations concerned with a single resource;

o   regional level organizations concerned with direct action for coastal resources conservation;

o   global organizations concerned with direct action for coastal resources conservation;

o   global organizations concerned with collecting, organizing, and sharing information to inform coastal management policy.

National Mangrove Committees (NATMANCOMs) which consist of qualified persons from government agencies, exemplify conservation NGOs concerned with a single resource. NATMANCOMs -- inspired by UNESCO -- have been established in India, Bangladesh, Thailand, Malaysia, Papua New Guinea, and Venezuela. These NATMANCOMs serve as (1) a communication link to UNESCO and other UN bodies; (2) an advisory group to government; (3) a coordinator of in-country research and training; and (4) a conservation watchdog. Since NATMANCOMs consist of a network of individuals who are influential in their respective professions, they have become effective spokespersons for coastal resource management in their nations.

Some NGOs have evolved from an adversarial posture to a collaborative arrangement with a national government. Members of the scientific community (including the British Royal Society) were instrumental in the re-designation of Aldabra Island from a proposed airstrip to a reserve. Now members of the royal society and Seychellois President France Albert Rene serve on the board of the Seychelles Island Foundation, whose mission is research and stewardship of protected areas (Stoddart and Ferrari, 1983).

6.2   Less Organized Actors

6.2.1   Subsistence and artisanal resource users.

The coastal zones of virtually all developing nations support some level of artisanal fishing, and other subsistence resource use or extraction. Mangrove forestry is both a source of subsistence and a commercial activity in many parts of Southeast Asia (Cragg, 1982; Snedaker and Getter, 1985). In parts of Africa and Asia, rice has been successfully cultivated for centuries on the landward fringe of the mangrove zone (Hamilton and Snedaker, 1984), and in India, mangrove forests support local honey gathering (Gentry, 1982). Subsistence-level resource users are often poorly organized, may have limited political influence, and may have little access to information about the interrelationships between coastal uses and environments.

Both national and international organizations may take steps to assist artisanal fishers, in effect creating a more formal organization. In Cape Verde, a government run company, the Sociedad Comercializado e apoio a Pesca Artesenal was created to assist the nation's 3,000 artisanal fishers.

6.2.2 Coastal landowners.

Another grouping of actors with important interests are coastal landowners and other coastal or inland residents. Often ownership of land is concentrated in the hands of relatively few owners. If national restrictions are imposed on coastal development to preserve "national heritage," or to ensure the right of all people to have access to the beach, individual landowners might see themselves as "losers", while the benefits of the coastal protection accrue very generally to other residents of the nearby coast or the hinterlands.

6.2.3 Ethnic groups.

Efforts to allocate coastal resources must often cope with longstanding conflicts among ethnic groups. The Indonesian land resettlement program (Koesoebiono, Collier and Burbridge, 1982), oil exploration in Malaysia (Klapp, 1984), and Sri Lankan coastal management (Amarasinghe and Wickremeratne, 1983) all exemplify conflicts among ethnic groups as well as competition among different economic sectors. Norton (1982) argues that:

> the intensity and pervasiveness of these conflicts arises from the compelling consciousness of social honor aroused by appeals to ethnic distinction and efficacy of this status ideology for organizing political solidarity and patronage.

Where ethnic groups disagree, simply adopting a regulatory program may not resolve underlying conflicts that could frustrate implementation of a new coastal management regime.

6.2.4 Social classes.

Virtually all social classes, from subsistence artisanal fishers, to middle class merchants and bureaucrats, to members of the aristocracy and the international jet set, depend on and affect the coastal zone. Social classes are not, strictly speaking, coastal actors, but they could be differentially affected by a coastal management regime. Gaining the involvement and cooperation of people from all income levels in integrated coastal management may be a complex task.

Social class, like ethnicity, may be envisioned as a sort of overlay on the other allegiances that coastal stakeholder users may have. In many nations, the strong separations between social classes could work against broad participation in setting coastal resource policy. This in turn may frustrate program implementation. Yet, the class system may be a deeply embedded part of the political culture, as it is in many Latin American nations. Without presupposing whether coastal management reinforces or helps rearrange prevailing distributions of income and power, it is clear that coastal managers must find ways to cope with the competing aspirations of classes. Communication among disparate groups is essential, although the precise forum and level of participation in decisionmaking will vary from nation to nation.

Negotiations within a class (i.e. agency heads, scientists, and members of the "aristocracy") may be possible, while negotiations across class lines may be much harder to arrange. Given these constraints, one might envision negotiations among ministries, state-owned enterprises, and major transnationals while artisanal fishers might be excluded. Often, social class reflects ethnic background, so dominant groups may exclude representatives of other classes and ethnic groups.

7. MANAGEMENT STRATEGIES

This section examines 11 distinct strategies for management of coastal resources and environments now in use throughout the developed and developing world. Our literature review has not identified any one document that defines and describes the full array of coastal management strategies. However, our review of the planning and environmental protection literature helped us identify 11 management strategies. Each of the strategies is a complex topic that could be dealt with at book length. For example, Biswas and Geping wrote a book on environmental impact assessment in developing nations (1987). Excellent reviews of marine protected area management are found in Salm and Clark (1985); and Silva and Desilvestre (1986). The book, **Managing Land-Use Conflicts**, describes the process of special area planning and presents eight case studies (Brower and Carol, 1987).

We do not assume that the management strategies we present constitute a definitive list. There could be more or fewer strategies depending on how one defines the terms "management strategy" and "coastal management."

Almost all developing nations are using two or more of the 11 management strategies identified below. We note that aside from regional seas, none of the management strategies is necessarily coastal zone specific. They have been used in inland as well as coastal areas. Shoreland exclusion, for example, has been used for managing inland rivers and lakes. The 11 strategies are:

- o **national economic planning;**

- o **broad-scope sectoral planning of coastal uses or resources;**

- o **regional seas;**

- o **nation- or state-wide land use planning and regulation;**

- o **special area or regional plans;**

- o **shoreland exclusion or restriction;**

- o **critical area protection;**

- o **environmental impact assessment of coastal development proposals;**

- o **mandatory policies and advisory guidelines;**

- o **acquisition programs;**

- o **coastal atlases and data banks.**

As Table 7.1 illustrates, these strategies are not mutually exclusive. In fact, they are usually mutually supportive. For example, France uses nation- or state-wide land use planning together with natural area acquisition campaigns as principal strategies. They are complemented by national economic planning, participation in the Regional Seas Program, special area planning, shoreland exclusion, and critical area protection. In the United States, all strategies are used to various degrees, with the exception of national economic planning and regional seas. Nation- and state-wide land use planning under a broad federal framework is the principal strategy in the United States. This strategy is reinforced with broad sectoral planning, special area and regional plans, impact assessment, acquisition programs and a coastal atlas and data bank. Shoreline exclusion and critical area protection are used less often.

Sri Lanka uses nation-wide land use planning and regulation as its principal strategy for coastal zone management supported by national economic planning and impact assessment. Indonesia depends heavily on environmental guidelines, reinforced with national economic planning, and is moving towards greater dependence on regional or specialized planning together with a national land use planning framework.

For each strategy, we define and describe the technique, cite examples of its use, and present important advantages and disadvantages. In cases where strategies are very similar, the distinctions are spelled out. The discussion also suggests which strategies are complementary in scope and purpose.

7.1 National Economic Planning

National economic planning involves setting prescriptive goals for each sector of the economy, affecting the allocation of labor, investment capital and land use. This style of planning occurs in both socialist countries and nations with a mix of central economic planning and private markets.

In some cases, planning decisions are centralized at the national levels; in others, targets for production are established at the regional level or through the intervention of central planning institutions and local authorities. The regional level is usually the prime focus for implementation of a national economic plan.

The main vehicle of national economic planning is usually a long-term plan, spanning a five year period. Production targets are set in most important sectors of the economy. Production, as the central feature of an economic plan, is then used to specify the size of the workforce, the type and quality of land needed for a particular industry, and the amount of investment capital needed to implement the plan. Besides striving to achieve production targets, national economic plans aim to affect a fast growth of the economy, reduce large disparities in income, and create employment.

Coastal sectors are fisheries, ports and shipping, transportation, agriculture, tourism, and industry. One potential strategy is to use national economic planning for the integration of sectors to produce an integrated coastal management program for a region. In this way, sectors such as fisheries, ports, and tourist development can be made mutually supportive. To some

54

Table 7.1: Strategies Used in Coastal Resource Management

	FRANCE	USA	GREECE	INDONESIA	SRI LANKA	THAILAND	PHILIPPINES
National economic Planning	X				X	X	X
Broad sectoral planning		X	X	X			
Regional seas	X		X	X			NA
Land use planning and regulation	O	O	O	X	O	O	O
Special area or regional plans	X	X	X	O		O	
Shoreline exclusion	X	SU	X	X		NA	X
Critical area protection	X	SU	X	X		X	X
Impact assessment		X	X	X	X	X	X
Policies and guidelines			X	O	X	X	X
Acquisition programs	O	O	X				
Coastal atlases - data banks		X	X	X	O	O	O

KEY:
O — Principal Strategy
X — Complementary Strategy
SU — Seldom Used
NA — Information Not Available

extent, the linkages between economic sectors will depend upon the resources available to the nation.

Second, recognizing the value of coastal resources in the economy leads to recognition of the impact of one sector on another. This in turn can foster a strategy of avoidance of unintended negative impacts.

The expansion of agriculture and mariculture in Indonesia at the expense of mangrove wetlands and estuarine habitats illustrates this point. In Indonesia, an Interagency Committee on National Policy and Planning for Coastal Zone Management provided policy ideas to the national planning agency to incorporate in the economic development plan for 1984 to 1989 (Kux, 1983).

National economic plans create a degree of certainty about the coastal frontage and adjacent land needed for development within a particular time frame. This avoids more random patterns of proposals and demands on coastal resources. This certainty, in turn, provides more time to accomplish the integration with other sectors and the avoidance of impacts described above.

However, five year plans may prove too rigid to take account of changing coastal circumstances. This rigidity may hamper a strong response to an environmental perturbation, such as the crash of a fishery. National economic plans have also been criticized for being too mechanistic and therefore obstructive to innovation, an effect that could also be felt in efforts to protect resources. At the other extreme, altering economic plans in response to every small perturbation in the economy is extremely disruptive for the agencies and productive units responsible for carrying out the plan.

7.2 Broad-scope Sectoral Planning

Traditional sectoral planning combines forecasting and implementation for capital investment, land use planning, and infrastructure needs for specific sectors of a national economy. Sectoral planning shares several characteristics with national economic planning (see Section 7.1), but places more emphasis on issues other than the production of economic goods.

Those sectors with greatest economic relevance to coastal management in developing nations are port planning, fisheries, and tourism. Given the close dependency of each of these sectors on a vigorous natural resource base, a consideration of habitat and environmental quality factors must be integrated with other aspects of sectoral planning to make the effort successful.

Several nations, recognizing the importance of environmental factors, have taken steps to include them in sectoral planning of a broader scope. In the United States, fishery plans for specific species prepared by the National Marine Fisheries Service (NMFS) under the U.S. Fisheries Conservation and Management Act are based on environmental system analyses that take into account sustainable yields, recruitment rates, water quality, and habitat quality.

The U.S. program also includes a capital investment dimension. Seed money has been granted to stimulate the organization of marketing cooperatives to help stabilize the economic fortunes of individual fishermen and stimulate fishery development of underutilized stocks. NMFS also tries to ensure that fishing does not interfere with other important marine resources. For example, they have worked with Gulf Coast shrimpers to try to avoid unnecessary mortality to endangered sea turtle species.

Given that most major ports are located in estuaries, port expansion is likely to preempt fringing wetlands, pollute water, and destroy productive benthic (bottom) communities. In addition, industrial facilities conflict spatially with public recreation or commercial fishing and preempt public access to the shore.

Japanese port authorities operating under the National Ports and Harbor Act and Ministry of Transport guidelines prepare comprehensive coastal management plans for their land and water jurisdiction (Inoue, 1984). The Japanese transportation sector in Japan is concerned with port modernization plans and urban development **and** environmental improvement programs. The process depicted by Figure 7.1 includes the preparation of an environmental impact statement and review of proposed plans by a local port council consisting of various interests including fisheries, recreation and citizens' groups (Inoue, 1984).

Successful tourism development requires a mix of attractive accommodations and shops, a suitable infrastructure (clean and sanitary water, good roads) and an accessible, relatively unspoiled environment. These goals can conflict with each other and with the development plans of other sectors.

Seventeen years ago Yugoslavia pioneered a program to balance tourism development on the Adriatic Coast, addressing both water supply and beach area use as well as maintenance of cultural values (Shankland and Cox, 1972). Oceanic island nations such as Western Samoa and Fiji recognize that tourism development must be planned in a manner that neither threatens the exceptional fragility of island ecosystems nor disrupts island societies (Towle, 1985).

In Brazil and Colombia broad sectoral plans have been completed for marine and coastal research. Both nations have established coordinating organizations linked to the national economic planning program to chart national programs for marine and coastal research. In Brazil, applied research objectives are directly related to information needs of mariculture development and estuary pollution control (Brazil, Comissao Intraministerial para os Recursos do Mar, 1980; Knecht, 1983).

Broad-scope sectoral planning represents a marginal change from the status quo. Since institutions tend to make only marginal adjustments when confronted with a need for change, broad-scope single sector planning is the most likely management strategy to be implemented. Broad-scope sectoral planning often serves as a transition to the integrated management strategies. If an agency broadens its horizons to assess the full range of impacts associated with its projects, and this wider perspective produces a net benefit to the agency, this positive experience should make the agency more amenable to taking the next step to an integrated management strategy.

The major disadvantage of broad-scope sectoral planning is the perpetuation of non-integrated, single purpose programs. Interest in integrated planning may be diverted by broad-scope sectoral planning, even though the

Figure 7.1: Procedure of Port Planning and Implementation in Japan

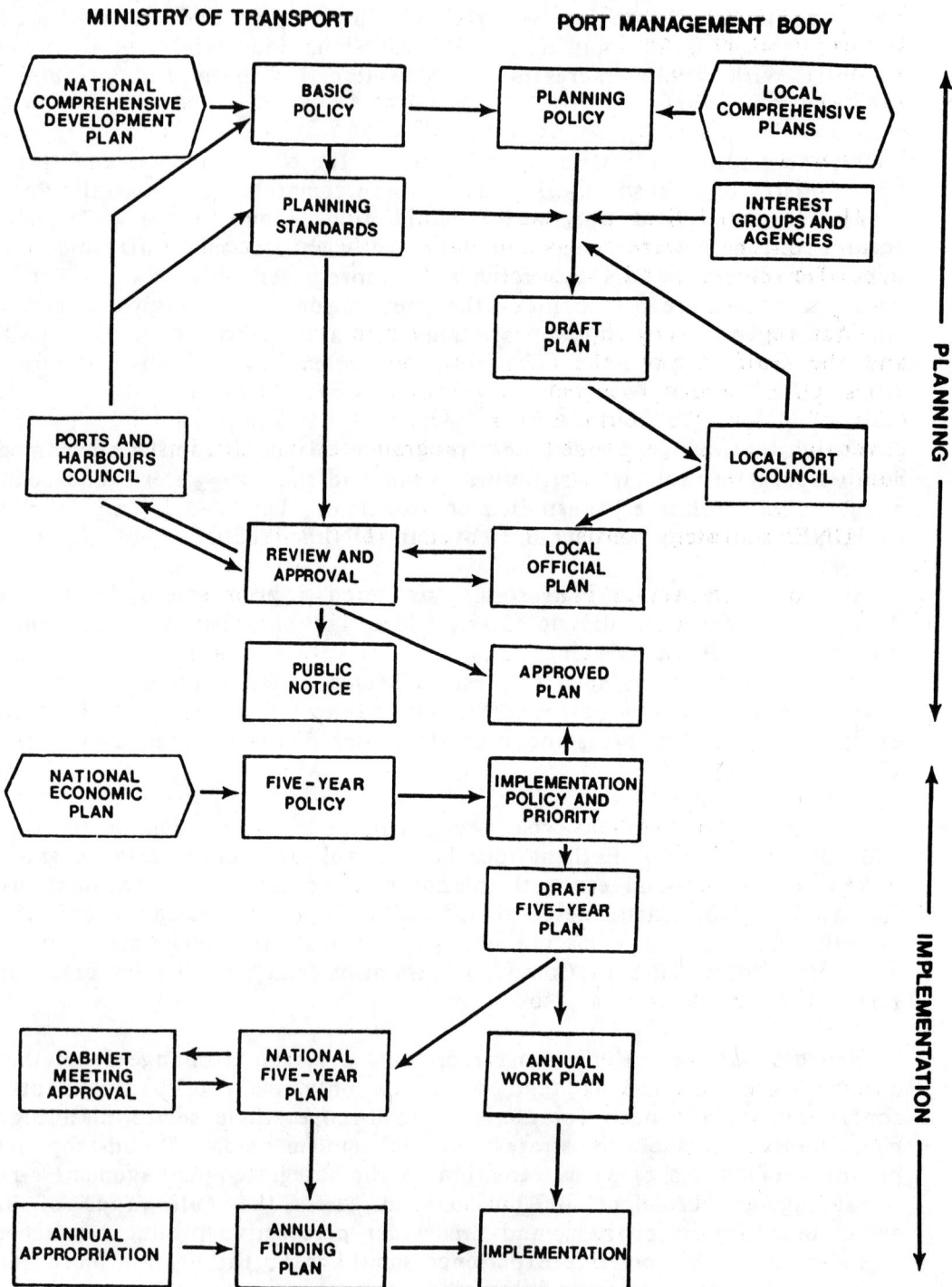

(Source: Inoue, 1984)

latter approach would be more effective in resolving an issue.

7.3 Regional Seas

The institutional parent of the Regional Seas Program is the United Nations Environment Program (UNEP). The program was created in late 1972, and is an outcome of the Stockholm Conference on the Human Environment (Hulm, 1983; UNEP, 1982b). UNEP's first governing council set the health of the oceans as its foremost concern in 1973, and it remains one of the leading issues today.

The Regional Seas Program was initiated by UNEP in 1974. The major ocean concerns were trans-boundary pollution, ocean dumping, fisheries, scientific research, and conservation.

Four regions were chosen for initial attention: the Mediterranean, Kuwait and the Gulf Region, the Caribbean, and West Africa. Over the next five years, UNEP added four more regions: the East Asian Seas, the Red Sea and Gulf of Aden, the South East Pacific, and the South Pacific. In 1980, the governing council expanded the program to include East Africa and the South-West Atlantic.

UNEP's strategy consists of four steps (UNEP, 1982b):

o   an Action Plan setting out activities for scientific research and cooperation, including assessment and management;

o   a legally-binding convention embodying general principles;

o   technical and specific protocols to deal with individual issues;

o   financial and institutional arrangements that implement the first three steps.

Each nation participating in a Regional Seas Program must adopt the Action Plan before the process can move forward.

Regional Seas Action Plans contain four parts: assessment, environmental management, legislation, and support measures (UNEP, 1982b). Assessment, the first priority, is geared to evaluate sources and effects of pollution, the state of living and marine resources, and development practices.

Management projects help build the capacity of local officials to make decisions and develop plans for coastal development. The legislative section includes regional conventions. Protocols may be adopted simultaneously, but often there is a significant time delay.

The Mediterranean Plan is a leading success of the Regional Seas Program. A companion agency to UNEP, the Food and Agriculture Organization (FAO), provided the early impetus which was broadened in 1975 when an action plan was adopted. The European Economic Community (EEC) has joined all Mediterranean nations, with the exception of Albania, in ratifying the convention. A "black list" of banned substances is identified by the anti-dumping protocol. The substances include mercury, cadmium, DDT, PCBs, radioactive wastes, some plastics, and lubricating oils. A third protocol against land-based substances was signed by twelve nations in Athens in 1980. Another protocol on protected areas, signed in 1980, is expected to increase the 15 marine parks located in the Mediterranean to a network of 100. Eighty-four marine laboratories participated in a first phase of water quality testing; a second phase will run until 1991.

By early 1983, some $8 million U.S. dollars were paid into a trust fund for Action Plan implementation by the 17 Mediterranean nations and the EEC. Offices oriented to specific aspects of the Action Plan are being opened around the region, consistent with UNEP's policy of delegating ultimate responsibility to the participating nations.

The voluntary participation of coastal nations in the Regional Seas Program helps to foster a sense of international goodwill, mutual benefit, and regional self-confidence. The program is flexible enough to allow nations and regions to concentrate on solutions that are most pressing, or for which there is already common agreement. In this way, a political momentum is generated to inspire efforts to address the more contentious issues. The requirement that all nations adopt the Action Plan and subsequent conventions and protocols helps to catalyze improvement in the environmental laws of developing nations. Another strength of this program is the explicit multilateral participation of scientists and scientific institutions.

A nation's participation in the Regional Seas Program may improve the nation's institutional capability, data base, and financial support for the following sets of transboundary issues:

o   marine pollution;

o   fisheries protection;

o   marine research of large-scale oceanographic phenomena (e.g., ocean currents, upswelling, or storm forecasting).

The Regional Seas mechanism may also help nations deal with other joint multinational interests, such as:

o   tourism (particularly in the Caribbean and the South Pacific);

o   mangrove conservation;

o   protection of migratory marine mammals and birds.

Since UNEP acts only at the request of national governments to formulate an action plan, the Regional Seas Program cannot respond quickly to resolve conflicts. Adoption of the Action Plan by all affected nations must precede further progress. The series of steps -- Action Plan, convention, protocol, implementation -- can take several years. (For example, the latest Mediterranean protocol to establish protected areas was signed five years after the Action Plan was adopted.) Since most members of Regional Seas Programs are developing nations, support measures for training, management and project implementation must be provided. Funding has been problematic, but UNEP has been fairly successful in "packaging" funds from UN multilateral and bilateral sources. As the name implies, members of Regional Seas are cognizant of major land-based pollution sources if they affect ocean quality. However, the program does not give special scrutiny to land use issues that impinge on coastal resources of a transboundary nature -- such as the conversion of mangrove ecosystems for maricultural and agricultural purposes and the consequent reduction of the regional shrimp fisheries.

7.4 Nation- or State-Wide Land Use Planning and Regulation

Land use planning and regulation specifies the type, intensity, and rate of development and conservation for a particular area. In this strategy the land use plans usually cover the entire coastal zone of the nation or state (or province). Broad goals and objectives are usually specified to direct the planning effort. Land use plans, consisting of both maps and policies, are usually translated into guidelines and legally-binding rules such as zoning ordinances. The earliest and still most common form of zoning is often called Euclidean zoning. This approach relies on assigning a single use designation (e.g., low density residential, central business district or heavy industry) to each parcel of land.

More recently, several variations on the theme have been proposed and implemented in some locations. Overlay zoning is often used to protect sensitive resources. This technique involves the imposition of special restrictions (e.g., requirements for setbacks or retention of wetland habitats) in addition to the designation of permitted land uses. Incentives may also be combined with land use designation by permitting greater density on coastal frontage to encourage development of high priority facilities, such as commercial fishing piers or other maritime commercial uses.

In the 34 countries that were members of the British Commonwealth, land use planning is termed Town and Country Planning. In the United Kingdom the Town and Country Planning Act requires local planning authorities to make careful surveys of their areas and to estimate needs over the next twenty years for housing, schools, industry, and roads (Waite, 1980). The authority then draws up a proposal showing how these needs will be met in land allocations, and prepares maps of various scales, depending on whether the subject is a town, a county borough or a county area. Town maps show proposed areas where special powers for land acquisition may be sought. The plan is supported by a written statement outlining the major proposals and a program map illustrating the phasing of development. A public hearing is held and an inspector -- a trained civil servant -- makes recommendations to the Minister for Town and Country Planning who decides whether or not to approve the plan. After plan approval, planning permission for all but relatively minor projects must be approved by local authorities. Local decisions may be appealed to the Minister.

In recent years, Town and Country Planning in England has delegated authority to local units of government. Additionally, public participation has been increased and administrative processes streamlined.

At least four countries have amended their Town and Country Planning programs to include a particular set of policies for land use control within a delineated coastal area (UNOETB, 1982a). The Bahamas prepared "development plans on an island-by-island basis, treating the coastal area as a separate planning entity." In Cyprus, under the Town and Country Planning Act "there are detailed regulations governing streets, construction and alteration of buildings . . . in coastal areas." Jamaica prepared a plan for coastal management which endorses "the evaluation of sensitivities and classification of areas of environmental concern." In Mauritius,

> the coastal area is regarded for physical planning purposes as both a separate entity and a part of national planning. The coast is dealt with as an entity in respect to recreation. For planning purposes coastal areas begin from 1 km inland of the high water mark to the end of the coral reef (about 50 m) (UNOETB, 1982a).

Guatemala may be another example of special policies for land use planning within a shoreland area. The coastal area of the country has been treated as a separate entity with regard to zoning. The coastal areas extend 3000 meters inland from the seashore (UNOETB, 1982a).

In the United States, California's requirement that all coastal cities and counties draw up a Local Coastal Program (LCP) is the basis for the most ambitious of the United States' coastal management programs with land use planning as its focus. Consisting of a land use plan and zoning regulations, an LCP must reflect the state policies on public access, water and marine resources, land environments, new development, ports, and energy facilities. Within the general framework of coastal resource protection, local governments have discretion over which goals to emphasize. The State Coastal Commission, the permit-letting agency for the coast, is responsible for reviewing all LCPs to ensure consistency with the policies embodied in the California Coastal Act. Following state approval, local governments are responsible for administration of the land use plan and implementation of zoning ordinances. However, the Coastal Commission still has oversight jurisdiction over sensitive habitats and areas immediately adjacent to the shoreline.

A state requirement that local governments prepare coastal zone land use plans is one of the most popular mechanisms used by coastal states to implement coastal zone management programs funded by the U.S. Coastal Zone Management Act. The arrangement has been termed state-local collaborative land use planning (Sorensen, 1978). To date, Alaska, California, Florida, Louisiana, Maine, Maryland, Michigan, Minnesota, North Carolina, Oregon, South Carolina, Washington, and Wisconsin have adopted the state-local collaborative land use planning model to implement their state coastal zone management program.

In Sweden the National Physical Plan empowered the government to define a national interest in a particular area, and order revision or preparation of a local master plan to address that national interest. Additionally, the government may order that the plan be legally-binding and declare a moratorium on development while the plan is pending. These steps were motivated by the finding that environmental quality was being impaired in many parts of the nation's coastal zone. Industrial siting, "holiday houses," and public recreation needs were key issues (Hildreth, 1975).

Ireland also follows the model of a central land use planning authority cast in the role of issuing guidelines for use by local land use authorities. The National Institute for Physical Planning and Construction Research has issued guidelines for use by local governments in amending Town and Country plans for coastal areas (Mitchell, 1982).

Although Nigeria does not actively operate a strong national coastal management program, it does have a Town and Country planning process. In March 1978, the federal government issued a land use decree which effectively nationalized all land not in productive use. "Theoretically, this makes possible comprehensive national regulation of development in presently unoccupied coastal areas" (Mitchell, 1982).

In Thailand, the government:

> has perceived the coastal zone to be an area important to the national economy, which can be extensively developed, especially in agriculture, fisheries, forestry, industry, tourism, and environmental conservation. As a result, many initial efforts to organize national coastal resource management programs have been developed. The Coastal Land Development Project was established in 1971 to facilitate proper planning of multiple use and management of coastal resources. Coastal land will be managed for eight types of use: coastal agriculture, fisheries, animal husbandry, salt farming, mangrove forest preservation, port construction, industrial zones, and tourism. When this plan is finished, it is to be submitted to the Coastal Land Development Committee for approval (Adhulavidhaya et al., 1982).

Land use planning presents a mechanism to resolve use conflicts arising either along the shoreline or at inland locations affecting the coasts. In this way the consequences of agriculture, watershed development and potential filling of wetlands can be addressed in the context of coastal resource management. When linked with strong zoning, land use planning provides clear guidance and certainty about future development -- both in pinpointing the precise location of future development and in specifying the types of uses allowed.

Programs organized at the state-wide or nation-wide level provide an opportunity to deal with multiple use conflicts in a consistent manner. Lessons learned in one local area can be adapted to other locations. Under both the Town and Country Planning and state-local collaborative land use planning arrangements described above, strong policies can be formulated by a central authority and adapted to local conditions, with an oversight role to ensure that the state or national interest is upheld.

63

Land use planning has often been criticized for its somewhat speculative nature. In most cases, plans are only as effective as the zoning ordinances and use restrictions that implement the plan. These in turn are guided by the integrity or political will of the government agencies responsible for plan administration. The principal vehicle for plan administration and implementation is the issuance of development permits. In most cases, land use plans alone cannot stimulate capital investment, nor do they ensure that development actually occurs on a specific plot of land.

For the developing countries, customs and traditions of opportunistic use of land combined with uncertain land tenure may complicate efforts to implement a clear, rational land use plan. This is most clearly evident in the proliferation of squatter settlements in most metropolitan areas.

Programs for land use planning instituted at the state or national level may override traditional local authority. Objections to this "preemption" have been registered in many locations, notably those U.S. states practicing state-local collaborations and Sweden. Local authorities in Sweden have objected to invasion by the national government of their traditional "planning monopoly" (Hildreth, 1975).

To be successful, land use planning requires an extensive information base consisting of data on a range of natural resource characteristics, historical settlement patterns, and institutional and political concerns. The strategy also requires the capability to interpret the data and fashion a single coherent plan. In most developing countries, the data base and the capability to synthesize the data may not be available. For example, in El Salvador a professional planner observed that approximately eighty percent of the essential information base is lacking. What little information exists is not adequate to support nation- or state-wide land use planning (UNOETB, 1982a).

Finally, land use planning does not provide a strong mechanism to cope with issues at the land/water interface. Neither does it address water column issues. Obvious examples here are protection of mangroves, coral reefs, submerged grass beds, or fisheries. The failure of land use planning to include water areas has been noted by a number of planners in developing nations (Baker, 1976; Beller, 1979; Amarasinghe and Wickremeratne, 1983; Mitchell, 1984).

7.5 Special Area or Regional Plans

"Special area plans" or "regional plans" refer to a coastal land or coastal resources use program that is larger than a local jurisdiction and smaller than an entire nation. The distinguishing characteristic of a special area or regional plan is the geographic coverage. The boundaries are usually delineated with two purposes in mind. First, they are intended to "capture" national resource or economic development issues that cross the boundaries of states or local governments. Such issues might include watershed management, protection of sensitive habitats, or development of a regional transportation network. Second, the boundary is drawn to encompass a significant natural resource, such as an embayment, river basin, estuary, mangrove hydrologic unit, or a littoral drift cell defined by shoreline erosion processes.

Special area or regional plans have a multi-sectoral perspective. They may focus on a single issue (such as tourism development), but interconnections are made with the other relevant sectors. Special area or regional plans are usually mandated by either a legislative body or a ministry of the nation or state.

The French government engages in large-scale coastal resource development programs acting through units of the special interministerial committee which oversees land use planning (Harrison and Sewell, 1979; France, Ministry of the Environment, 1980). The principal tools of these units are extensive legal powers and substantial budgets for planning and capital works construction. The coasts of Languedoc and the Aquitaine region are the focus of efforts to attract visitors from other heavily used resort areas. Some of the early work in the Languedoc region generated significant environmental impacts. A chain of resorts extends along 125 miles of coast. Roads, hotels, and marinas were installed, wetlands filled, harbors and lakes deepened, and artificial beaches created. In contrast, the planning for the Aquitaine region has been recognized as a model of sensitive coastal development and conservation (Mitchell, 1982).

In Greece, the most successful part of the national coastal effort has been the development of regional coastal planning programs "which have contributed to the mobilization of regional and national interests for cooperation in [resolving] coastal issues" (Camhis and Coccossis, 1982). Crete has been one of the most active areas. The Chaind region of Eastern Crete was elected as a pilot project for an in-depth examination of coastal management issues. The purpose of this effort, besides solution of problems in the area, was to provide information for program evaluation. The pilot program especially emphasized developing an appropriate implementation strategy with the cooperation of local authorities and the public (Camhis and Coccossis, 1982).

Indonesia's program for integrated coastal swamplands development in Sumatra presents a good example of the regional planning approach in a developing nation (Hanson and Koesoebiono, 1979). The principal goal of this effort was to reduce the number of inhabitants in the densely populated regions, which are also the most productive lands for agriculture. The government policy was relocation in relatively underdeveloped areas in Sumatra. However, most of these areas were of marginal value for intensive agriculture, such as rice cultivation. This forced the government to make difficult choices between settlements in erodible uplands and settlement in estuarine deltas with productive forests and shrimp fisheries.

Given the large number of questions about the optimal use and management of marginal lands in Sumatra, an integrated program was suggested for area development and environmental management. Between 1969 and 1974, six pilot projects were developed by the Ministry of Public Works (P.U.T.L.). These trial efforts led to a commitment in 1974 to open one million hectares of delta lowland. Together with the finding that rice crops could be grown within one year after swampland is directly connected to a river, an awareness grew of the need to pay close attention to coastal zone environmental factors (Hanson and Koesoebiono, 1979).

The Ministry of Public Work's **Tidal Swamp Reclamation - The Second Five Year Development Plan 1974/75 - 1978/79** urged that development planning take account of environmental factors and that resources be managed on a sustained-yield basis. Throughout the country, regional planning units have been set up at the provincial and county level. A growing interest in impact assessment in Indonesia has led University scientists to prepare tables and matrices to evaluate likely consequences of development action and to transmit them to decision makers. Planning is complicated by two systems of land tenure. One arises out of **adat** law in which resource rights are vested in village units; the second arises from national law. Further, Indonesian decision making involves multiple agencies with poorly defined channels of authority. Hence, decisions are often reached by a gradual process of consensus (Hanson and Koesoebiono, 1979).

Some analysts have argued for a more formal set of agency responsibilities to carry out the conceptually sound regional planning approach (Hanson and Koesoebiono, 1979). It was recommended that The Regional Planning Office (BAPPEDA) became a focal point for impact assessment. A specialized unit for resource management and protection was also recommended, to coordinate the three agencies which have coastal zone responsibility,.

In the United States, one of the first regional planning bodies with an effective implementation program was the San Francisco Bay Conservation and Development Commission (BCDC). The agency was created in 1965 in response to citizen and legislative concern over the alarming rate of peripheral filling of San Francisco Bay and the consequent shrinkage of the Bay's size. Initially the agency was endowed with limited permit granting authority for the Bay shoreline and directed to report to the legislature on long-term regulatory needs. The outcome was a Bay Plan and permanent mandate to approve or deny projects that would fill bay bottoms or block public access within 100 feet of the shore -- a policy which still exists. Decisions are made by a commission comprised of a mix of local governments, agencies and citizens. BCDC has virtually halted the net loss of wetland acreage and bay bottoms, yet has permitted construction of needed port and airport facilities along the Bay's edge by obtaining mitigation in the form of wetland restoration (Swanson, 1975).

Australia established the Port Phillip Authority in 1966 to cope with major issues confronting the state of Victoria -- notably coastal erosion, land-use conflicts, and lack of coordination between public agencies (Cullen, 1977; Cullen, 1982). Membership of the Authority is drawn from a mix of public representatives and the pre-existing agencies concerned with aspects of coastal development: the Departments of Crown Lands and Survey, Public Works, two local government representatives, and two citizen representatives. Jurisdiction extends 200 meters landward and 600 meters seaward. Responsibilities include coordinating development in the Port Phillip area, preserving existing beauty and preventing deterioration of the foreshore, and improving facilities for public use in the Port Phillip area. The Authority was strengthened in 1980 to improve its permit enforcement capabilities. Based on the general success of the Port Phillip Authority, the Victorian government later extended many aspects of the approach to the rest of the state (Cullen, 1982). Recently the Port Phillip Authority was terminated in order to form a comprehensive coastal management program for the entire state of Victoria (Victoria, 1988).

The regional level of planning and analysis confers a number of advantages which are absent from local and/or national level planning. At the regional level, it is possible to address and resolve resource issues confronting whole ecosystems, such as siltation of an estuary as a result of development in its watershed. Very often these issues cross a number of jurisdictions and cannot be dealt with effectively without a regional geographic focus. Coastal management institutions organized at the regional level -- like BCDC and the Port Philip Authority -- often present an opportunity for local government authorities and officials with responsibility in various sectors affecting the region to cooperate and resolve common problems. This trend was illustrated by the examples drawn from Australia, California, and Indonesia.

By choosing a regional focus, national governments are able to concentrate on the areas with the most pressing problems. At the same time, a regional program can serve as a model which can be tested, modified and perhaps extended to other regions. This was the approach used in Greece, Australia, and California.

Most regional planning exercises have a predominantly landward focus and do not explicitly deal with water-based issues such as fisheries management. Either the regional agency does not have the regulatory authority for water areas and resources, or it chooses not to exercise its authority in the "wet side of the coastal zone." BCDC, the California agency, has no significant authority to manage bay fisheries. Similarly, the French planning exercises are predominantly land use planning linked with capital works and resort development (Harrison and Sewell, 1979). The Indonesian example, though still at an early level of development, offers some promise to deal more specifically with the "wet side" fisheries and wetland habitats.

The issue of local autonomy may arise during the creation of a regional planning agency by a state- or national-level legislative body. This opposition is the most significant where traditions of local government control are strong. Consider the case in which a politically powerful city (such as the national capitol and largest metropolitan area) is situated on an estuary. If that estuary is the object of a nationally-sanctioned regional planning effort, conflicts could arise over the economic or port development goals of the city and the regional need to preserve fishery, mariculture, tourism, and recreation resources.

7.6 Shoreland Exclusion or Restriction

Shoreline exclusion or restriction refers to regulatory programs that prohibit or significantly limit certain uses within a strip or band in the coastal zone. The areas subject to shoreline restriction are typically landward of the high water mark; they are rarely the intertidal zone or submerged lands because the national government usually controls those areas under separate mandates. In developing nations the shoreline exclusion strategy commonly arises from three concerns: blockage of public access, degradation of views, and erosion of shorelines. Residential development and tourist development appear to be the primary targets of shoreline exclusion. In some cases, exclusion zones and land use planning boundaries for permit letting are mutually supportive and may be integrated into a single program. Shoreline exclusion zones differ from critical area management programs (see Section 7.7), in that they are coast-wide and do not carry a special designation declaring the uniqueness of particular types of areas.

67

There are two types of shoreland exclusion programs: (1) those with fixed upland and offshore dimensions; and (2) those based on the features of the shoreland. Shoreland exclusion zones vary in size. Figure 7.2 displays the inland extent of exclusion zones in 22 nations or states. The 22 examples we identified are illustrative, and are not a definitive listing of this management strategy. The extent of the zones depicted in Figure 7.2 vary from eight meters to three kilometers.

The concept of public ownership of land along the shoreline is a historical tradition in many countries. In Australia and New Zealand, a shoreland zone of Crown Lands constitutes this public area. In both countries tradition has not been maintained in some areas, and Crown Lands have been sold off ("alienated") for development. Some steps have been taken, however, to preserve the public interest in Crown Lands through shoreline exclusions. Since 1851, coastal subdividers in New Zealand have been required to set aside esplanade or foreshore reserves for public open space uses. These foreshores consist of strips of land 66 feet wide paralleling the mean high tide line (Chapman, 1974).

In Latin America at least eight countries apply the concept of a zona publica -- or public zone. Figure 7.2 indicates that Brazil, Chile, Colombia, Costa Rica, Ecuador, Mexico, Uruguay and Venezuela have established shoreland zones based on a specific setback from the shoreline (usually mean high tide). The figure depicts the considerable variation among these seven countries in the width of the zona publica. There are also considerable international differences in both the uses that are allowed in the zona publica and the extent to which the government forcefully plans and manages the area. (Sorensen, forthcoming). For example, in Uruguay the 250 meter zone is designed to preserve natural resources and foster tourism, but only the mining of sand and other beach materials is expressly prohibited by the law (Calvo, 1988).

Costa Rica appears to have one of the most ambitious and comprehensive shoreland restriction programs in the world (Sorensen, 1990). The jurisdictional area is a 200 meter wide marine and terrestrial zone. The law divides the zone into two components: the "zona publica" and the "zona restringida" (restricted zone). The zona publica extends inland 50 meters from mean high tide or the inland limit of the wetlands and the upstream limit of the estuaries as defined by salt or tidal influence. The zona restringida covers the remaining 150 meters inland. The zona publica is devoted to public use and access, and commercial development is generally prohibited. Exceptions to the prohibition against commercial development are made for enterprises that are coastal-dependent, such as sport fishing installations, port installations, and their infrastructure. In the zona restringida, development is controlled by a permit and concession system that is based on a detailed regulation plan formulated at the local level of government. A concession is a development right on a specific parcel of land for a particular land use and fixed time period.

In Greece, the National Coastal Management Program imposes "strict controls" within a 500 meter band on both sides of the shoreline. Greece departs from the general pattern and imposes controls both landward and seaward of mean high tide (Camhis and Coccossis, 1982).

Figure 7.2: Shoreland Exclusion or Restriction Setbacks

COUNTRIES	DISTANCE INLAND FROM SHORELINE*
Ecuador	– 8 m.
Hawaii	-- 40 ft.
Philippines (mangrove greenbelt)	----- 20 m.
Mexico	----- 20 m.
Brazil	------ 33 m.
New Zealand	------- 66 ft.
Oregon	------------ Permanent vegetation line (variable)
Colombia	------------------ 50 m.
Costa Rica (public zone)	------------------ 50 m.
Indonesia**	------------------ 50 m.
Venezuela	------------------ 50 m.
Chile	-------------------- 80 m.
France	---------------------- 100 m.
Norway (no building)	---------------------- 100 m.
Sweden (no building)	---------------------- 100 m. (in some places to 300 m.)
Spain	---------------------- 100 to 200 m.
Costa Rica (restricted zone)	------- 50 m. to ----------- 200 m.
Uruguay	----------------------------- 250 m.
Indonesia** (mangrove greenbelt)	------------------------------------- 400 m.
Greece	--- 500 m.
Denmark (no summer homes)	--- 1-3 km.
USSR - Coast of the Black Sea (exclusion of new factories)	-------------------------- 3 km.

* Definition of shoreline varies, but it is usually the mean high tide. Most nations and states exempt coastal dependent installations such as harbor developments and marinas.

**Indonesia has both a 50 m setback for forest cutting and a 400 m "greenbelt" for fishery support purposes (see text for explanation).

In the United States, a program to protect resources and guarantee public access in the state of Oregon exemplifies a coast-wide exclusion based on the configuration of natural features. A state supreme court decision upheld a century-old law requiring that the entire foredune area (to the inland line of permanent vegetation) be kept free of construction and fencing to ensure the continued right of access. This restriction also confers the benefit of protecting dune vegetation and associated wildlife (Oregon, 1976).

The Bahamas offer another example of a shoreland exclusion determined by the characteristics of the site. The government's **Planning Guidelines for the Control of Land Use and Development in the Commonwealth of Bahamas** (Bahamas, 1976) do not fix the setback requirement for building in the coastal area, but require "a view from the sea," whereas, in the city, the limit is set by distance (from 15 to 30 feet) from the street (UNOETB, 1982a).

Norway has both a fixed and a variable setback. No building is allowed within the first 100 meters, and second (vacation) home development is set back as far as necessary to control the adverse effects of residential construction (UNOETB, 1982a). Denmark has a similar exclusion program for beach protection. Its Conservation of Nature Act provides protection against construction and landscape changes in a 100 meter coastal zone. Guidelines drawn up by the Ministry of Environment prohibit building of summer houses within a protected belt of one to three kilometers from the coast (UNOETB, 1982a).

Shoreland exclusion or restriction programs are administratively attractive: they are inexpensive, geographically precise, and offer clear guidance about prohibited uses. This administrative simplicity provides a high degree of certainty for both coastal management agencies and potential developers. Such zones can be tailored to particular natural resource features such as dunes, mangroves, or other wetland habitats, to ensure that they are protected wherever they occur in the coastal zone. Exclusion programs providing a setback for public access and shoreline recreation are likely to enjoy wide support from inland residents who don't own coastal property. Shoreline exclusion zones with specific dimensions provide consistency throughout a nation or subnational unit. In a situation where coastal resources are being degraded at an alarming rate, exclusion or restriction zones are a convenient way to impose a moratorium on development until a more comprehensive land use plan can be prepared and implemented.

This technique would complement a broader program of coastal land use planning for a state or nation. Exclusion or restriction zones can both constitute the first thrust for the declaration of public trust and form the core of a permanent system to limit modification where sensitive resources occur, as well as allow development elsewhere on a permit basis.

In developing nations, the concept of the public right to gain access to and along the coastline may be a persuasive political argument for coastal zone management. Costa Rica exemplifies this situation. Exclusions providing a setback for public access and shoreline recreation are likely to enjoy wide support from inland residents.

Highly developed or urbanized coasts present difficult or impossible circumstances for the use of exclusion zones. Imposition of an exclusion zone would be opposed in political circumstances where native citizen private property owners have enjoyed a high degree of discretion in implementing their own development plans. Similarly, it is doubtful that exclusion programs could be adopted without strong support from a nation's legislative body or the chief executive.

The inland exclusion distance often is not great enough to address the issues the strategy was established to resolve. For instance, public access (or view protection) may not be guaranteed by any boundary line that is seaward of the public road nearest the coast. Similarly, effective control of shore erosion hazards may not be achieved unless the exclusion zone includes the entire shore area that can be expected to erode during the lifetime of existing or proposed development (e.g., 50-75 years). Without a complementary program of land use planning or some other effective planning strategy (e.g., sectoral planning), exclusion programs alone leave large gaps in a national effort to achieve an integrated coastal management program based on a coastal systems perspective, as outlined in Section 2.6.

7.7 Critical Area Protection

Critical area protection programs are enacted by state or national governments to achieve one or more of the following purposes: (1) to conserve or preserve a particular type of sensitive environment or natural area (such as mangroves, wetlands, barrier islands, coral reefs, and endangered species habitats), (2) to preclude development on selected eroding coasts, or (3) to restrict development in a special flood plain. In the context of the first purpose, critical area protection is very similar to sectoral planning for wildlife protection. In the second and third contexts -- hazard protection -- critical area protection is very similar to exclusion zones.

Three features distinguish critical area protection as a management strategy. First, **a formal designation sets the stage for the program.** Often this is a result of an inventory of resources and a screening of candidate sites, and a recommendation from an agency staff person to a decision making body. Second, critical area programs are **not implemented on a coast-wide basis --** such as for all of the nation's mangrove forests. Instead, they are **selected for specific geographic locations --** such as the mangrove forests bordering Guayaquil Bay. Third, designated critical areas typically **address the concerns of more than one sector;** they simultaneously serve the purposes of wildlife protection, hazard area management, parks, and perhaps research. The strategy of critical area designations often represents an intermediate step before the creation of wildlife refuges, parks, or hazard control zones. Area designations for eroding coasts may precede a shoreland exclusion strategy.

The International Union for the Conservation of Nature and Natural Resources (IUCN) has assisted nations to establish protected areas for habitat preservation and conservation of genetic diversity. Both marine and terrestrial areas adjacent to the shoreline are included in IUCN's global system of reserves.

A commonly used technique to implement critical area programs is to severely restrict development, usually in perpetuity. The mechanisms to ensure that no development occurs may include some form of purchase, ministerial

71

restriction, or condemnation. Often an activities plan is prepared for the delineated critical area in order to prevent use conflicts. In some cases, education or research programs are organized to take advantage of the resources in the critical area (McNeeley and Miller, 1983).

Land use planning for a "buffer zone" around the core resource area is sometimes incorporated in critical area programs. In other cases, an environmental assessment must precede any project in or adjacent to the critical area. Under Indonesia's National Forestry Act, a 50 meter wide belt of "protection forest" must be maintained along coastlines for mangrove silviculture and a 20 meter wide belt must be kept intact along river banks. A complementary program involves reforesting upland areas to promote the goal of sustainable yield. A more speculative prospect is the reconversion of marginally productive agricultural land to mangroves (UNOETB, 1982a).

In Queensland, shoreline erosion is addressed through the Beach Protection Authority (BPA), which maintains a 400 meter jurisdiction. Within this jurisdiction, special Beach Erosion Control Districts have been created, within which no development may proceed without BPA's approval. BPA can also control sand removal or vehicle use within a Beach Erosion Control District (Cullen, 1982).

The Barbados Parks and Beaches Commission Act prepares regulations governing beach protection, sanitary conditions and practices for managing public parks and beaches (UNOETB, 1982a).

Critical area management shows promise as a technique to help developing nations avoid the consequences of urbanization in flood plains and agriculture or forestry on steep erodible slopes, two common problems world-wide. This management strategy enables a nation to concentrate funds and staff resources on the most threatened or hazard-prone areas of the coastal zone. The very term "critical area" alerts citizens and decision makers to the need for quick action. Since many designated areas support more than one important resource or hazard, the critical area strategy provides the flexibility to tailor a detailed site plan or management approach to unique local conditions. Often this is preferable to routine use of general environmental guidelines (see Section 7.9).

The designation strategy can also be used as a stop gap measure until a more programmatic solution can be found through shoreland exclusion or perhaps a more standard sectoral plan for parks, research or erosion control. Administration is relatively simple, and overall costs are low.

Critical area designation, like acquisition programs (see Section 7.10), is seldom a complete response to a resource issue. It is likely to be more comprehensive than acquisition alone, however, because critical area protection usually has both a land use regulatory program and rules for guiding human activities within the area. A critical area designation may, however, become the focal point of intense political controversy.

7.8 Environmental Impact Assessment

A notable outcome of the Stockholm Conference on the Environment was the international diffusion of environmental impact assessment ("EIA"). The term is used to describe both a governmental process and an analytic method. As a process, EIA is usually imposed by government to force public agencies -- and in some cases private developers -- to disclose environmental impacts, to coordinate aspects of planning, and to submit development proposals for review. As an analytic method, EIA is used to predict the effects of a project or a program. The three fundamental premises underlying environmental impact assessment are:

o   cause and effect relationships can be determined
     with reasonable accuracy and presented in terms
     understood by policy makers;

o   prediction of impacts will improve planning and
     decision making;

o   the government can enforce decisions emanating
     from the impact assessment process.

The EIA process includes assessment of a proposed project's potential effects on the sustained use of renewable coastal resources as well as the potential effects on the quality of the human environment. The process is mandated by law or executive decree and generally involves a procedure that requires the following information: **(1) the characteristics of the project site; (2) a description of the project; and (3) a description of the consequences or impacts of a project for different dimensions of the environment.** Usually it also requires that alternatives to the project be identified and comparatively assessed and spells out measures to avoid or mitigate impacts.

Typically, the procedure of impact identification and assessment of its severity is combined with an institutional process requiring preparation of a formal document or holding of a hearing in order to describe environmental impacts and strategies to reduce them. A specific agency or ministry is given responsibility for being the focal point of the EIA process. The outcome of this process is often the imposition of mitigation measures as a condition of project execution. These measures may take the form of design changes, shifts in project location, or changes in the order in which different portions of the project are constructed. For example, a resort development may be redesigned to avoid destruction of dune vegetation and prevent interruption with a natural sand supply, or construction of a pier may be timed to avoid interference with the spawning cycle of a commercially important fish.

In this discussion, it has been noted that impact assessment is usually focused on the project level. Assessments may also focus at the program level, such as for a river basin development initiative. This programmatic approach is less common in developing countries (Horberry, 1983). Program level assessment -- when done for a large geographic area -- is conceptually similar to regional planning but does not include a mechanism to compel actual plan making and implementation.

Three different standards of review are commonly used to decide whether an EIA is needed. Depending on the agency and its mandate, the EIA process may be invoked as follows:

o   for all projects in the coastal zone or other sensitive areas (Greece's coastal management program has this requirement);

o   for any project likely to create significant environmental impact (the most common situation);

o   for any project of a specific type (e.g. major roads, large public works projects).

In the United States, the National Environmental Policy Act (NEPA) and subsequent guidelines established the legal framework for impact assessment. Its requirement that the environmental consequences of federal projects, **and their alternatives**, be assessed extends to all U.S. supported international activities with potentially significant adverse environmental impacts, including projects funded by USAID. Special emphasis is to be placed on irreversible impacts and on the cumulative effect of a project together with past and future projects.

The World Bank also requires that environmental impacts of projects be assessed. Most other regional development banks have recognized that good investment policy requires an accounting of projects in terms of both economic and environmental feasibility. In fact, failure to consider environmental impacts has been cited as the cause of major shortcomings in the success of resource development projects -- particularly large-scale impoundments (World Environment Report, 1982).

In the Netherlands, procedures for environmental impact assessment have been regarded as a significant contribution to that nation's coastal management effort (Wiggerts and Koekebakker, 1982). Greece requires an impact statement on all projects within five kilometers of the shore (Camhis and Coccossis, 1982). In Sri Lanka, the environmental impact assessment is an integral part of their coastal zone management program. The director of the Coast Conservation Department has the discretion to require a developer to submit an impact assessment. Figure 7.3 shows the impact assessment procedure in Sri Lanka.

The European Community has proposed a Directive on Environmental Impact Assessment that, once ratified, would bind all member states. The proposal calls for developers of certain types of projects to submit an EIA when seeking project approval. An open process is suggested, requiring the responsible agency to coordinate with other agencies and to make the report public before rendering a decision (Camhis and Coccossis, 1982).

Developing nations with assessment requirements include Brazil, Thailand, the Philippines, Indonesia, and India. Since Sri Lanka requires an environmental impact statement for all major development (Amarasinghe and Wickremeratne, 1983), USAID funded an environmental assessment in that country for a major irrigation program of the multi-donor Mahaweli Development Program. The study evaluated land use changes, losses to forestry and wildlife, soil erosion, water quality changes, reduction of wetlands, effects on fisheries, and several social issues. Extensive recommendations were made on natural system management which were carried forward in an action plan with special emphasis on animal migration corridors

Figure 7 . 3: Sri Lanka Coast Conservation Department's (CCD) Procedure for Reviewing and Issuing Permits

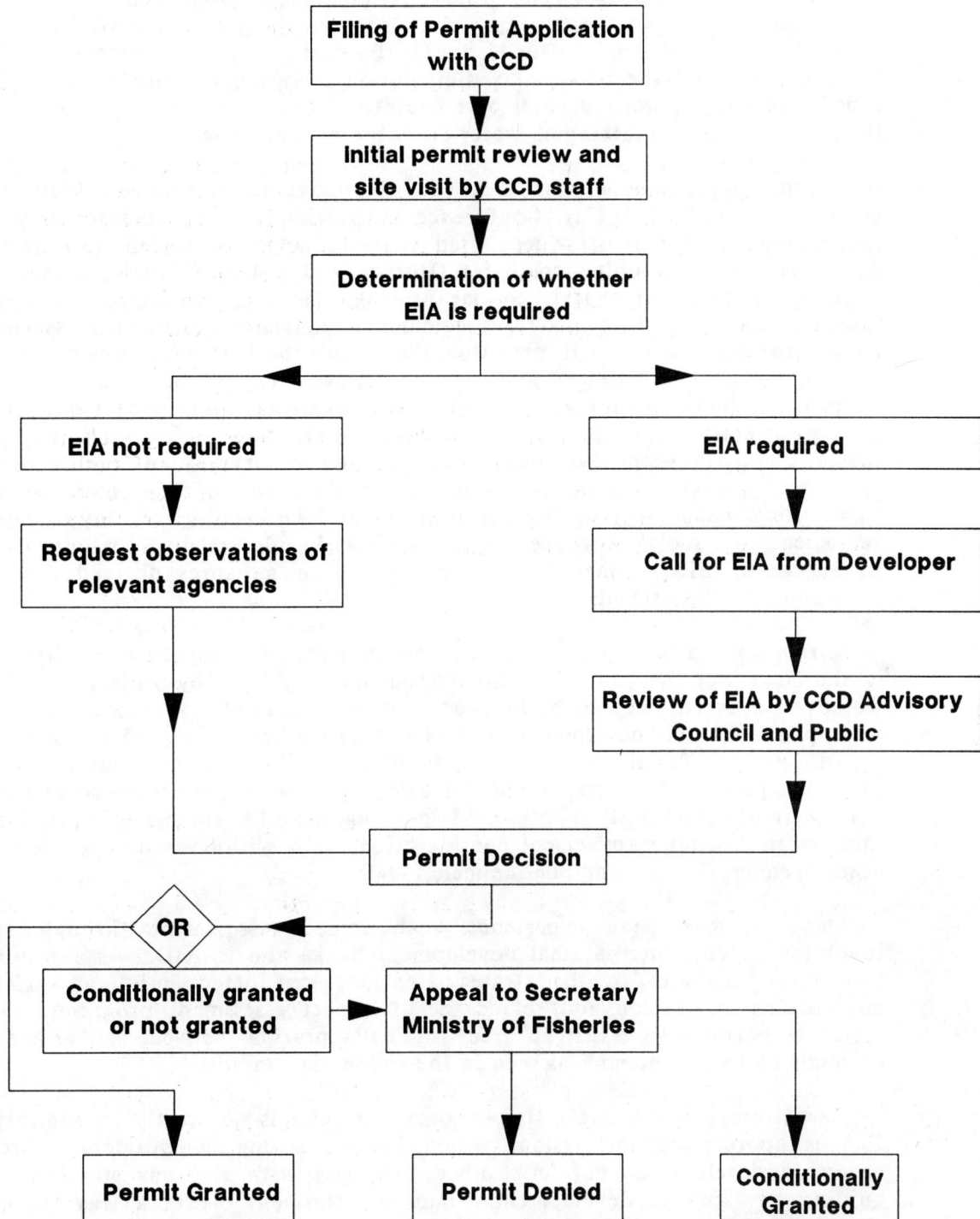

```
                    ┌────────────────────────────┐
                    │ Filing of Permit Application│
                    │        with CCD            │
                    └────────────────────────────┘
                                  │
                                  ▼
                    ┌────────────────────────────┐
                    │  Initial permit review and │
                    │   site visit by CCD staff  │
                    └────────────────────────────┘
                                  │
                                  ▼
                    ┌────────────────────────────┐
                    │  Determination of whether  │
                    │      EIA is required       │
                    └────────────────────────────┘
                                  │
          ┌───────────────────────┴────────────────────────┐
          ▼                                                 ▼
┌──────────────────────┐                    ┌──────────────────────┐
│   EIA not required   │                    │    EIA required      │
└──────────────────────┘                    └──────────────────────┘
          │                                             │
          ▼                                             ▼
┌──────────────────────┐                    ┌──────────────────────┐
│ Request observations │                    │ Call for EIA from    │
│  of relevant agencies│                    │     Developer        │
└──────────────────────┘                    └──────────────────────┘
          │                                             │
          │                                             ▼
          │                               ┌──────────────────────────┐
          │                               │ Review of EIA by CCD      │
          │                               │ Advisory Council and Public│
          │                               └──────────────────────────┘
          │                                             │
          ▼             ┌────────────────────┐          │
          └────────────▶│  Permit Decision   │◀─────────┘
                        └────────────────────┘
                                  │
                            ◇ OR ◇
          ┌───────────────────────┘
          ▼
┌──────────────────────┐        ┌──────────────────────┐
│ Conditionally granted│───────▶│ Appeal to Secretary  │
│    or not granted    │        │ Ministry of Fisheries│
└──────────────────────┘        └──────────────────────┘
          │                       │                  │
          ▼                       ▼                  ▼
┌──────────────────┐   ┌──────────────────┐  ┌──────────────────┐
│  Permit Granted  │   │  Permit Denied   │  │  Conditionally   │
│                  │   │                  │  │     Granted      │
└──────────────────┘   └──────────────────┘  └──────────────────┘
```

(Source: Sri Lanka, 1988)

and wildlife conservation. Horberry (1983) cites the case as unusual because the impact assessment was carried forward into a specific environmental planning program implemented by local authorities.

The UNEP Regional Office for Asia and the Pacific reported that an Environmental Impact Assessment on a deep sea port near the outlet of the Songkhla Lake basin influenced final port design (Horberry, 1983). Documentation of the potential sedimentation of the lake from dredging and construction and pollution from port operations caused a change in the site of the port to minimize mixing of water from the port and the lake.

Given the impetus of the U.S. 1969 National Environmental Policy Act (NEPA) and the 1972 U.N. Conference in Stockholm, impact assessment has become one of the most widely used coastal management strategies in both developed and developing countries (Horberry, 1984). This wide usage and relatively long-term experience means that the methodology for impact assessment is well developed and commonly understood. Impact assessment procedures produce better information about both the host environment and the project, and serve to define and separate issues. Unlike broad-scale sectoral planning, regional planning or national economic planning, impact assessment focuses attention on the details of projects that cause use conflicts in the coastal zone. Program level impact assessment can serve as an "early warning system" to avert the worst consequences of large-scale efforts such as river basin development plans and ensure that the ecological, hydrological, geological, and social consequences are adequately addressed. Finally, a major advantage of EIA is that it provides mitigation measures derived from an environmental assessment.

Mitigation planning in a region, or the pooling of mitigation requirements, is the next step beyond EIAs carried out on a project-by-project basis and would be a useful way to build good integrated coastal resource management principles into the development sector. Thailand's National Environmental Board (NEB), the administrator of the nation's EIA program, has developed extensive guidelines on environmental assessments with particular reference to coastal areas. The NEB is interested in using the EIA process to expand and incorporate coastal management considerations both within its own agency and other agencies (Kinsey and Sondheimer, 1984).

Many nations have experience with impact assessment through their involvement with international development banks and USAID. As a result of this early exposure to the strategy, nations can often build on existing mechanisms to develop and refine useful impact assessment programs. This seems a particularly fruitful area for collaboration between governmental officials and academicians, as seen in the Indonesia example.

The strategy is relatively simple to execute and is not costly to administer. EIA is appropriate for nations which have a strong commitment to rapid economic development but lack other strategies with rigorous standards for guiding new coastal development. Impact assessment offers a way to make changes in project design and location, thus avoiding the most serious use conflicts without undermining the attractiveness of a project in economic or social terms.

The most common objection to environmental impact assessments is that they are only information reports or "report cards"; they are not decision documents. Their effect typically occurs late in the development process so they accomplish only minor, and perhaps insignificant "fine tuning" of project location or design. Conversely, impact assessment has been criticized for putting roadblocks in the way of timely project completion. These concerns are most likely to arise where institutions invoke impact assessment as an afterthought, or "add on", rather than an integral part of the planning process.

The identification and assessment of potential impacts is only as good as the available data base. Experience in developing countries suggests that the amount and quality of data is steadily improving but is still deficient in many, if not most, areas. Collaboration of universities and government agencies may be one way to overcome this deficiency.

Environmental impact assessment is fundamentally an analytic and interpretive procedure; it is not a substitute for sound policies. Without a clear, straightforward translation of an assessment into a specific action such as a change in project design, the EIA strategy is usually not meaningful but only a cosmetic exercise.

A further vexing problem in EIA is the difficulty of assessing the cumulative effects of environmental alteration. Impact assessment is most often conducted on an ad hoc or project-by-project basis. Few agencies have found suitable procedures to predict and account for impacts of a series of projects in a particular region or ecosystem over a period of time. A related problem is the difficulty in identifying thresholds -- levels of change beyond which irreversible damage occurs.

7.9 Mandatory Policies and Advisory Guidelines

Many -- if not most -- state or national coastal management plans are based on a set of policies and guidelines. As the adjectives imply, the distinction between policies and guidelines is that the former requires compliance and the latter is voluntary. In other words, the text of a policy can be identical to the text of a guideline. The difference is often only the verb "shall" or "should". The institutional arrangement for voluntary guidelines is also usually different from the arrangement for implementing mandatory policies. Policies are used by government units that have the power to issue permits and prepare specific land use plans. Advisory guidelines are issued by government units that do not have authority for implementation. These government units must depend on other government units to apply their guidelines in their permit letting or plan making activities.

Policies and guidelines are formulated to provide a framework for issuing permits as well as preparing land use or special area plans. These management strategies were described in Sections 7.4 and 7.5. Policies and guidelines usually precede the preparation of land use and special area plans. Unlike land use plans and special area plans, policies and guidelines do not refer to a specific geographic locations. If they did they would be land use or special area plans. Policies and guidelines are usually organized according to types of uses (e.g., tourism development, channel dredging, spoil disposal); or types of environments (e.g., wetlands, mangroves).

The coastal plans of California and Sri Lanka are two good examples of mandatory policies. The law that established the California Coastal Zone Conservation Commissions in 1972 required submission of a plan to the legislature in 1976. The enabling legislation provided very little guidance on either the organization or the composition of a coastal plan. In December 1975 the California Commissions published a 443 page plan. One hundred and sixty two policies are the core of the **California Coastal Plan** (California Coastal Zone Conservation Commissions, 1975). The policies cover every scale and type of coastal development and conservation -- from the heating swimming pools to the siting of nuclear power plants. A set of regional summaries and 1:125,000 scale maps for the entire coastline of California are included in the Plan. The maps illustrate "the location and extent of coastal resources, developed areas and other factors that influence coastal planning".

Although it was possible to get some idea of where many of the 162 policies would be applied, the **California Coastal Plan** was widely criticized for being too vague. No one could read the set of policies, guidelines and maps and predict how their particular interest in the coast would be affected. The Plan proposed that predicability would be provided by each of the 15 coastal counties and 54 cities preparing a local coastal program (see Section 7.4). Also, the Coastal Commissions would continue to issue permits and use the Plan's policies as its reference. When the local coastal programs were approved by the Commissions, most of the permit letting authority would be relinquished to the respective local governments.

The law creating a coastal zone management program in Sri Lanka was modeled on the California program. Like its model, the Sri Lanka Coastal Zone Management Plan produced in 1987 is also a policy plan (Sri Lanka Coast Conservation Department, 1988). There is little in the Sri Lanka Plan that is geographically referenced. The policies are organized into three groups: types of environments (e.g., estuaries, corals, mangroves, dunes); types of resources (e.g., archaeologic, historical, scenic sites); and administrative procedures.

The major problem with policy plans is the uncertainty they create or perpetuate. In both the California and Sri Lanka coastal zone management plans it is difficult to tell how one's interests will be affected by the policies. However, uncertainty does have its advantages. Policy plans create less opposition from pro-development interests than land use or special area plans. No one in California can tell for sure how the policies will be applied to their coastal land or resource. During the debate on the 1976 law to implement the **California Coastal Plan** one particular phrase captured this dilemma: "Is it better to be vague and insidious or specific and outrageous?" If the Plan had been a land use plan for the entire coastal zone, most of the people with an economic interest in land or resources in the zone would have coalesced into a large and vocal opposition force.

Advisory guidelines are usually multisectoral in scope; they may address a range of project types and natural resources and social and cultural issues. In this respect, they differ from broad-scope sectoral planning, a management strategy which incorporates environmental considerations into planning for a single sector of a nation's economy (Section 7.2). Guidelines are also similar to the model of land use planning in which central authorities draft guidelines for incorporation in plans prepared at the local level. However, the guideline strategy, by definition, does not mandate preparation of a specific plan or implementation measures.

Adoption of national guidelines is exemplified by the joint efforts of the Indonesian National Committee on the Environment, the Indonesian Institute of Sciences, and other leading universities to prepare "General Guidelines on the Development and Management of Coastal Areas" (ASEAN, 1983). The management guidelines were organized as follows:

o   inventory of natural resources;

o   human settlement;

o   land use and development allocation;

o   environmental considerations in project planning and the development of coastal resources;

o   food production and raw materials;

o   conservation and environmental protection;

o   recreation and tourism;

o   infrastructures and engineering works;

o   construction materials;

o   public health;

o   management of water resources;

o   institutional framework;

o   navigation, shipping and harbors;

o   security.

A review copy of the guidelines was circulated to a variety of departments and used for six years. The Office of the Minister for Development Supervision and the Environment (the successor to earlier environment agencies) plans to revise the document to reflect both users' comments and environmental laws. The book may also be translated into English for review or use by ASEAN member countries (ASEAN, 1983).

An opposite response to the adoption of guidelines was exemplified by Ecuador in 1981. At that time the nation considered, but did not adopt, coastal development and conservation guidelines. It was concluded that on a nation-wide basis, conditions in the coastal zone and along the continental shelf varied too greatly to apply uniform guidelines (Vallejo, 1987).

International assistance agencies have produced a considerable number of guidelines for types of projects and environments -- many of which have direct or indirect bearing on coastal management. For example, UNEP produced a pamphlet on "Coastal Tourism" (Ahmed, 1982) as part of its environmental guidelines series. Similarly, the UNOETB (1982b) produced a manual on technologies for coastal erosion.

Guidelines can serve a valuable educational function. They can offer general direction for project design and construction and raise the level of awareness and understanding among agency and government staff. Drafting and revising guidelines also serves as a vehicle for intergovernmental communication as well as a forum for government agencies and interest groups concerned with coastal management, as exemplified by the Indonesian experience. Guidelines can sensitize planners and policy makers in different sectoral or functional divisions to issues that require horizontal or vertical integration of government efforts. They have also been shown to be of assistance to any private sector that has an interest in development within the coastal zone. The guidelines should act as a handbook to provide foreknowledge of the government's policies and concerns regarding the impacts the proposal may generate. In some cases, advisory guidelines contain hidden power because of the strength or influence of the agency issuing them. The perceived threat of formal imposition of guidelines by law may inspire voluntary compliance by developers.

A survey of 92 environmental guidelines publications produced by international assistance institutions reached several conclusions relevant to coastal resource management in developing nations:

> The fact that we found so little evidence of the systematic application of the existing guidelines suggests that either they have been tried and found useless or that agencies have not made sufficient resources and incentives available to sustain their use. We suggest that some agencies never put some guidelines into operation because their function is to improve public relations or to provide educational material to the development community in general. In other cases, staff of agencies do not use guidelines systematically because the guidance is too general or incompatible with real tasks and problems. In many cases, staff do not use guidelines because agencies do not require their use, nor provide the appropriate training and resources, nor establish any institutional penalties for failing to use them (Horberry, 1983).

7.10 Acquisition Programs

We refer to an acquisition program as an organized effort over several years for systematic land purchase, which is distinct from a one-time acquisition effort. In developed nations, acquisition is usually the single most reliable way to secure the future of a sensitive resource or to ensure that land is available for a specific type of development for public use, such as a port facility or a park. Acquisition programs may be carried out by the public sector, non-government organizations dedicated to particular resource protection or development purposes, or a partnership of public and private sectors.

In capitalist developed nations and some middle income developing nations, acquisition of specific parcels often represents the final implementation of a critical area protection program, as exemplified by the U.S. estuarine research reserve program. It may also be used to implement portions of a land use plan or as implementation for sectoral planning for parks and reserves.

France operates an acquisition program to implement "a land policy of coastal protection respecting the natural landscape and ecological balance." The Coastal Conservatoire is empowered to acquire land using preemption in cases anticipated by law or through appropriation. In addition, the Conservatoire is allowed to receive legacies and donations and may enter into covenants with individuals to secure protection of the shoreline.

Policies and priorities for land acquisition are set by the Conservatoire's Administrative Council, a 34 member body comprised of elected officials and representatives of agencies and associations. Directions for action are based on the information of seven shoreline councils: North Sea, Atlantic-Bretagne, Mediterranean, Corsica, lakes, French shores of America, and French shores of the Indian Ocean. The Conservatoire can intervene in any shoreline community of more than 100 hectares. A report indicated that 65 sites had been acquired representing 10,000 hectares and 120 kilometers of coast. Goals set in 1980 called for acquisition of 50,000 hectares in the following years (France, Ministry of the Environment, 1980).

British experience with acquisition to achieve coastal protection dates back to 1895, when the National Trust for Places of Historic Interest and National Beauty was formed. A private organization, the Trust accomplished its first acquisition on coastal cliffs at Dinas Oleu, near Balmouth, Wales. Other individual acquisitions followed and in 1962-1963, the Trust inventoried the coast to identify suitable sites. In 1965, Enterprise Neptune was launched -- a campaign for fund-raising and coastal acquisition. The government opened the fund-raising with a 250,000 pound contribution. In two and a half years the fund grew to 1 million pounds, with private contributions, and stood at 2 million pounds by 1978. By 1976, 333 miles of coast had been saved by acquisition or covenant. Stewardship activities complement the land purchase work of the Trust (Steers, 1978). England's National Trust program has served as a model for similar citizen-oriented efforts in New Zealand and Japan (Chapman, 1974; Shapiro, 1984a).

Acquisition is seldom a complete response to a significant coastal resource issue. There may be an erroneous tendency to assume that a problem is solved once an acquisition transaction is complete. For example, a land acquisition program for important wetland habitats can be frustrated by poor land use practices in the surrounding watershed, causing excessive siltation in the wetland basin. Beyond the problem of managing adjacent land uses, the acquisition must be followed up by a vigorous program of stewardship to ensure that the initial acquisition objective is fulfilled. This may mean monitoring easements or covenants to guarantee a free, well-signed public right-of-way, or a well-protected endangered species habitat.

Though administratively cost-effective, acquisition is certainly one of the most expensive coastal management strategies. Unless a sustained flow of funds can be assured and earmarked for exclusive use in acquisitions, this strategy is not likely to prove effective.

At this time, major acquisition programs for coastal protection appear to be concentrated in developed countries. As development pressure begins to impinge on the most sensitive resources in developing nations, acquisition campaigns are likely to become more important.

The financial, legal, and administrative costs of acquisition programs should be kept in mind by developing coastal nations. Many developing nations have relatively rural, agricultural, or vacant shorelines. The generally undeveloped nature of their shorelines -- particularly when combined with liberal constitutional provisions for either the taking of land or restriction of private property development -- provides the opportunity for ensuring public use of shorelands, hazard control, and resources conservation through exclusion or critical area strategies. Nigeria's nationalization of all land not in productive use and Costa Rica's creation of an exclusion zone both illustrate the relative ease with which many developing nations may impose restrictions or acquire private property without full compensation to property owners. Obviously, a developing nation's imposition of the exclusion zone or the critical area strategy is far less costly than the acquisition strategy developed nations are often forced to use as the only option available to achieve the same coastal zone management objectives.

7.11 Coastal Atlases and Data Banks

A coastal atlas and data bank are systematic compilation, interpretation, and display of information linked to a specific set of coastal issues, organized for an entire state or nation. The premise of coastal atlases is described in a document prepared by the State of Texas:

> Through inventory and evaluation of coastal zone resources, environments, and land and water uses, programs can be established that will permit use of natural resources and maintenance of environmental quality by adjusting use to resource capacity (Brown et al., 1980).

Although simple data or mapping for one site or several sites can assist the policy making process, such an effort is not regarded as an atlas or comprehensive data base. Rather, several features must be present to qualify an information system as an atlas or data bank:

o   information collected should be issue-oriented, designed to lay the foundation for policy making;

o   information should be collected consistently for the same parameters, and preferably at the same scale -- on a nation- or state-wide basis;

o   information should be compiled and synthesized in meaningful ways, using consistent weighing and scaling techniques;

o   information should be easily retrievable.

A coastal atlas meets the criteria outlined above and, in addition, includes a reproducible set of maps prepared on a common scale. In some cases, the map may represent the final output of the data base. In other cases, preparation of a series of descriptive and interpretive maps may be part of the analytic effort. For example, an initial round of maps might be prepared to delineate biological, geographical and land use features on a stretch of coast. Next, a second round of maps may be prepared. At this stage, a map of slope stability could be prepared using maps of geological units, slope, and historical landslides. At the third stage, a composite map of all geologic hazards could be compiled, indicating levels of risk for new development and indicating areas to be avoided.

The same approach could be used to combine maps of shellfish beds, wetlands, and endangered species habitats into a single map of sensitive biological resources. The resulting maps would give planners and policy makers tools to guide the type and intensity of new development, or to choose priority areas for protection or acquisition.

For coastal zone management purposes, a data base refers to a set of information systematically organized around consistent geographic units. For example, a data base could be keyed to parcels or townships of land, an offshore tract, or a particular linear kilometer of coastline. Often the data base is conceptually organized as a table with information on a set of natural resource parameters (geologic material, soil type, vegetation cover, prevailing land use, agricultural suitability) keyed to each geographic unit. Alternatively, a coastal pollution data base might be organized as a network of points reflecting the location of monitoring stations for water quality. With the advent of reliable, low-cost computer automation, there is a pronounced trend towards computer storage of data bases. This, in turn, allows easy updating of information and completion of a variety of computations.

Several U.S. states have prepared state-wide atlases of their coasts as the information foundation for their coastal management program. Florida launched a mapping effort in the early 1970's, and Texas followed a few years later. One of the more ambitious efforts was completed by the state of Washington in collaboration with the University of Washington's Geography Department. Over 30 parameters are mapped for each coastal county, each keyed to policies regulating shoreline development.

The European Commission recognized the need for consistent reliable mapped data and in 1973 recommended a program be initiated for "classifying the territory of the community on the basis of environmental characteristics" (Briggs and Hansom, 1982). The role of "Ecological Mapping" in the coastal zone was reiterated in the European Coastal Charter (Briggs and Hansom, 1982). Although a case study was carried out for the Basilicata area of Italy, the proposed method does not evaluate the coastal zone as a separate entity. Four specific applications of data base and coastal stages have been suggested for the European Community: flood hazard mapping, erosion hazard mapping, coastal pollution, and landscape and habitat evaluation (Briggs and Hansom, 1982).

The Philippines' Coastal Zone Program has undertaken a program of data collection for selected areas via analysis of LANDSAT images (Zamora, 1979). A national survey of coastal resource use is under way and a four volume report has been prepared. Approximately ten years ago, the Japanese government collected 24 natural and social factors pertaining to the coastal zone (Shapiro, 1984b). The data was computer-mapped for a band extending one kilometer on either side of the shoreline. A coastal atlas was prepared for Osaka Bay at a scale of 1:25,000 (Shapiro, 1984b). The Osaka atlas was prepared by university students, faculty, and citizens groups to influence the government's coastal development policy making process.

Sri Lanka is preparing maps of the coastal zone with technical assistance funded by USAID. Much of the work is being completed by students and faculty of the Geography Department, Peridynia University (USAID, 1982). New Zealand has compiled an **Atlas of Coastal Resources**. The announcement for the Atlas proclaims:

> It will be of interest to all those who use the coast to work and play, and of particular value to students and teachers, engineers, planners, scientists, fishermen, boat owners, divers, marine farmers, and many others (Tortell, 1981).

The Eastern Caribbean Natural Areas Management Program (ECNAMP), a non-governmental organization, has assisted in the preparation of a series of island areas in the Eastern Caribbean. That effort drew heavily on the skills and capabilities of island residents and included an integral training component (Towle, 1985).

Coastal zone atlases and data bases can play a central role in facilitating a more integrated and better informed approach to coastal resource management. These strategies promote sound organization of the often fragmented information existing for the coast. By drawing together data from different aspects of the environment -- on mangrove location, shrimp production, and land use designations, for example -- data bases emphasize the interaction of specific components of the environment. Often a coastal atlas or data bank is first used as a tool for problem identification, perhaps directing attention to sites that need immediate attention.

Coastal atlas and data bank preparation has direct connections to the Regional Seas Program for those nations that border on constricted ocean areas. If the coastal zone issues are transboundary in nature, data banks and atlas programs may have to be designed for two or more nations if the products are to be effectively applied. Regional preparation of an atlas or data bank should also realize savings to be achieved by economies of scale. To be effective as a management tool, as distinct from a problem identification technique, coastal atlases and data bases must be linked to a prescriptive set of policies and actions based on the assembled information.

Like the strategies of impact assessment and acquisition, a coastal atlas can yield valuable educational benefits. The educational benefits are derived not only from the product, but also from the compilation process. This is especially true if an open process is used involving all relevant government agencies and non-governmental organizations. If the product is presented in a clear, attractive format, maps of the coastal zone can also help convey the need for regulation, acquisition, or capital investment. This in turn can help generate support for coastal management policies among citizens, interest groups, agency personnel, and elected officials.

Since atlases and data bases record the condition of the coast at a given moment in time, they provide a valuable benchmark to be used as the basis for future comparisons. In this way, rates and patterns of natural changes can be measured and the effectiveness of a particular regulatory program can be evaluated. Computerized data banks are especially suited to periodic updating for tracking progress. A second technology that advances the case of atlases and data banks is LANDSAT imagery, which is ideal at a gross scale for preparing base maps and identifying resources and generates new data at frequent intervals. Since academicians can often make valuable contributions to data bases, a nation adopting this approach is likely to benefit from collaboration between universities and environmental agencies.

The utility of coastal zone atlases and data banks is governed by several constraints. First, these strategies are fundamentally tools for compilation and synthesis of information. They must be linked to a process of interpretation of findings, policy setting and intervention in the form of regulation acquisition or capital investment and construction to be considered a management strategy. Many initial attempts to build atlases and data bases are not linked to a specific policy making process which spells out how the findings are to be applied. Without setting clear goals for the relationship between data collection and implementation, nations that prepare atlases and data bases may be disappointed with the result. It is common for the information assembled to have only marginal application to the policy making questions asked. By contrast, the environmental impact assessment strategy is tied to the analytic process by formal institutional procedures for report preparation or project revision.

Second, it is clear that the value of a coastal data base or atlas is critically dependent on the quality and quantity of raw information. In developing countries, the available data is often uneven with regard to accuracy and consistency of coverage. Third, the methods by which data is compiled, scaled and aggregated have an equal impact on the utility of the data base or atlas. This is especially evident in considering the map scales at which data is obtained and reproduced. For instance, maps compiled at 1:250,000 or 1:125,000 are useful for large-scale regional planning, but much finer grain is needed (perhaps 1:24,000) for preparation of land use plans. Even more detailed maps are needed for site plans of particular projects. Fourth, atlases and data bases can quickly become obsolete, so there must be a commitment to their timely use and continual updating. Finally, building an atlas or data base is costly in dollar and staff terms. It should not be undertaken without a clear realization of both start up **and** maintenance costs. Since the methods, contents and results derived from coastal zone atlases and data banks vary so widely, systematic evaluation of these techniques should be undertaken.

NOTE: Permit letting. One of the questions most frequently raised by the first edition of this book was: "Why isn't permit letting listed as a management strategy?" Our response is that permit letting is an integral component of at least six of the 11 management strategies. Land use planning, special area or regional plans, shoreland exclusion, critical area protection, environmental impact assessment, and mandatory policies and advisory guidelines are all undertaken to provide government units with policies and information for making decisions on the issuance of the permits that are required (usually by law) before proposed development actions can proceed to construction or implementation. In most cases, permit letting presents a government unit with four options. The permit can be issued with, or without, any conditions attached to the development proposal. Similarly a permit can be denied with, or without, conditions attached to the resubmission of the development proposal.

In both developed and developing nations, the government unit that carries out the management strategy often does not have permit letting authority to ensure that its decisions are adequately implemented. In this situation the government unit is relatively powerless and can only affect decision making by making recommendations to the government units that do issue permits. This is the case in the United States and Thailand with respect to environmental impact assessment and impact statements. In Sri Lanka the impact assessment law was amended in 1988 to give the Central Environment Authority the power to require impact statements and deny proposals (Baldwin, 1988). It is clear that the potential for adequately implementing a management strategy is greatly increased if the governmental unit also has permit letting authority to back up its decisions.

8. GOVERNANCE ARRANGEMENTS

This chapter presents important concepts related to the development of governance arrangements and reviews the major institutional arrangements nations have used to manage their coasts. We begin by examining the complexity of national governance arrangements and explaining the need for better sectoral integration. Section 8.3 gives an overview of institutional arrangements and supplements. Some of the major choices nations face in creating an institutional arrangement are outlined in Section 8.4. Major institutional arrangements to broaden the scope of sectoral planning are described in Section 8.5. Then, Section 8.6 explains several supplements to traditional institutional arrangements.

We define **institutional arrangement** as **the composite of laws, customs, and organizations established by society to allocate scarce resources and competing values.** Every coastal nation has established its own institutional arrangement for managing coastal resources and environments. Five important components of a society's institutional arrangement are:

  o   legal and administrative authorities;

  o   customs and traditions;

  o   governance arrangements;

  o   non-governmental organizations;

  o   management strategies.

Society creates institutional arrangements to allocate scarce resources and to resolve conflicts among competing values. Accordingly, it is appropriate to ask whether there are similar institutional arrangements across nations for the resolution of coastal issues. If there are common types of arrangements, are some more efficient or equitable in the way they help resolve coastal issues? We describe the major types of governance arrangements in this chapter to enable a comparison of institutional arrangements across nations.

8.1   Complexity of the National Governance Arrangement

As countries increase their level of socio-economic development, their arrangements for national governance become more complex. There are three sources of complexity. These are **sectoral, functional,** and **hierarchical** differentiation. We explain these three factors to set the stage for a discussion of alternative arrangements for managing the coast.

8.1.1 Sectoral and functional differentiation.

As governments specialize in a discrete policy area, sectoral differentiation is the result. Chapter 2, coastal management definitions, mentioned sectoral differentiation in the context of sectoral management or planning. In the realm of coastal management, specialization tends to focus on policy areas formed by coastal uses (e.g. fisheries, ports and harbors, water supply and wastewater disposal, and tourism). Table 8.1 indicates that 15 to 25 sectors may be represented in a nations' coastal zone. Often, each of the sectors listed in Table 8.1 has at least one responsible government agency. The numerous sectoral divisions and corresponding government bureaucracies that affect coastal uses, resources, and environments complicate coastal management. (Other social policy areas such as criminal justice or public education involve fewer government sectors.)

The potential for fragmentation of governmental responsibility and duplication of effort increases with the number of sectoral divisions in a policy area such as coastal management. Figure 8.1 indicates that "horizontal integration" describes efforts to coordinate the separate sectoral divisions and thereby reduce fragmentation and duplication.

Each governmental sector is also divided by functional specialization and differentiation. Functional divisions provide the structure for governmental intervention. Examples which commonly occur in the governance of coastal resources and environments are:

   o   generating and disseminating information (including
        research and education);

   o   levying charges;

   o   taxing;

   o   funding and/or constructing projects and programs;

   o   acquiring, managing, and selling property;

   o   long-range policy-setting and planning;

   o   regulating private development and operations,
        particularly permit letting.

These divisions are listed in ascending order according to the relative degree of government intervention. Regulation is the greatest degree of intervention.

Functional division of government sectors tends to create separate agencies. For example, in Ecuador two different government units set fishing policy, another two units are responsible for fisheries research, another unit administers technical training for fishing, and still another unit funds fishing enterprises (Vallejo and Caparro, 1981).

The differentiation and specialization of functions in each sector thus increases the potential for fragmentation of responsibility and duplication of effort. Figure 8.1 indicates that vertical integration attempts to coordinate the separate functional divisions of a government sector, such as fisheries. The combination of functional and sectoral differentiation produces a matrix arrangement of government organizations as illustrated by Figure 8.1. The

Table 8.1: Sectoral Planning and Development in the Coastal Zone

Sectors that are often coastal zone or ocean specific	Sectors that are rarely coastal zone specific but have direct impacts
1. Navy and other national defense operations (e.g. testing, coastguard, customs)	1. Agriculture - Mariculture
	2. Forestry
2. Port and harbor development (including shipping channels)	3. Fish and wildlife management
	4. Parks and recreation
3. Shipping and navigation	5. Education
4. Recreational boating and harbors	6. Public health - mosquito control & food
5. Commercial and recreational fishing	7. Housing
	8. Water pollution control
6. Mariculture	9. Water supply
7. Tourism	10. Transportation
8. Marine and coastal research	11. Flood control
9. Shoreline erosion control	12. Oil and gas development
	13. Mining
	14. Industrial development
	15. Energy generation

Figure 8 . 1: The Arrangement of Government Organization in the United States for Selected Sectors

Functions (functional division)	Port Devel- opment	Fisheries	Pollution Control	Parks & Recreation	Marine Research
Generate and Disseminate Information	UNI MA LPD	NMF SFG UNI	EPA SWQ UNI	NPS SPD FS UNI LG	NSF UNI FWS NOA ONR
Levy Charges	LPD	NMF SFG	SWQ LG	NPS LG FS SPD	NOT APPLICABLE
Taxation	IRS LPD STB	IRS STB SFG	IRS STB	IRS STB	TAX EXEMPT
Fund &/ or Construct Projects & Programs	COE EDA LPD	NMF EDA SBA	EPA SWQ LG	NPS SPD FS LG	NSF NMF FWS UNI ONR
Acquire, Manage & Sell Property	GSA CG LPD	GSA LG	GSA LG	NPS LG FS SPD	GSA NOA UNI FWS
Policy Setting & Plan Making	CG LPD COE MA	NMF SFG SD	EPA SWQ COE CG	NPS SPD FS LG LG	NSF UNI FWS NMF ONR NOA
Regulation (Permit Letting)	CG LPD COE EPA FWS	NMF SFG CG	EPA COE SWQ LG	NPS SPD LG FS	NOA FWS SFG

Vertical Integration

← **HORIZONTAL INTEGRATION** →

ABBREVIATION KEY:

CG	U.S. Coast Guard
COE	U.S. Army Corps of Engineers
EDA	U.S. Economic Development Administration
EPA	U.S. Environmental Protection Agency
FS	U.S. Forest Service
FWS	U.S. Fish and Wildlife Service
GSA	U.S. General Services Administration
IRS	U.S. Internal Revenue Service
LG	Local Government (City or County)
LPD	Local Port District
MA	U.S. Maritime Administration
NMF	National Marine Fisheries Service
NOA	National Oceanic & Atmospheric Administration
NPS	National Park Service
NSF	National Science Foundation
ONR	Office of Naval Research
SBA	U.S. Small Business Administration
SD	U.S. State Department
SFG	State Fish and Game Department
SPD	State Park Department
STB	State Tax Board
SWQ	State Water Quality Agency
UNI	Universities and Colleges

matrix depicts the basic complexity of a nation's governance arrangement for integrated coastal zone management. Agencies with involvement in each of the sectoral functions are shown in the boxes produced by the intersection of the two sides of the matrix. The agencies indicated in Figure 8.1 are only a few examples drawn from U.S. national, state, and local government involvement in coastal management.

If a coastal nation has 25 sectoral divisions that directly or indirectly influence coastal uses and resources (such as those listed in Table 8.1) and seven functional governmental divisions govern each of these sectors (as illustrated by Figure 8.1), the product is **one hundred and seventy-five** separate points of potential government involvement for a fully integrated coastal resources management program. In practice, however, not all functions are performed in each sector. An analysis and inventory of U.S. national government involvement in coastal resources identified 83 different federal units of government with responsibilities that affected coastal zone uses and resources (Gamman, Towers, and Sorensen, 1974), half that of the hypothetical total.

The matrix illustrates that an agency often performs many functions for a particular sector (vertical integration). The Environmental Protection Agency (EPA), for instance, was created to integrate all aspects of pollution control. It is also common for one governmental unit to have the same or different functional responsibilities across more than one sector (horizontal integration). The matrix shows that the U.S. Army Corps of Engineers (COE) has responsibilities in both the port development and pollution control sectors. These consolidations of agency responsibility reduce the number of different government units with functional responsibilities in sectors that affect coastal management. However, the consolidations are often more than offset by governmental tendencies to further subdivide sectoral functions into both geographic jurisdictions (such as regions) and activity jurisdictions (such as dividing the regulation of port development into dredging permits and waste water discharge permits).

8.1.2 Geographic and activity subdivisions.

A number of geographic subdivisions of sectoral functions in the United States can be seen in Figure 8.1. For example, the National Park Service (NPS) provides recreation within the jurisdictional boundaries of its coastal parks and the Forest Service (FS) provides recreation within the boundaries of coastal forests. Figure 8.2 depicts the geographical division of New Zealand's coastal zone prior to the 1990 Resource Management Law Reform (RMLR). The RMLR is expected to streamline and consolidate the divisions illustrated by Figure 8.2. Regulation of fresh water and coastal pollution (under the Water and Soil Conservation Act of 1967) is the responsibility of the Ministry of Works and Development. However, the geographic extent of this function stops at the oceanward limit of territorial waters (three nautical miles). Coastal pollution oceanward from this boundary to the outer limit of the fishing zone (five nautical miles) is the responsibility of the Ministry of Transportation.

The geographical division of the same sectoral function is a vexing problem in coastal zone management, as Figures 8.1, 8.2 and 8.3 illustrate. The

Figure 8.2: Legislative Administration of the Coastal Environment of New Zealand

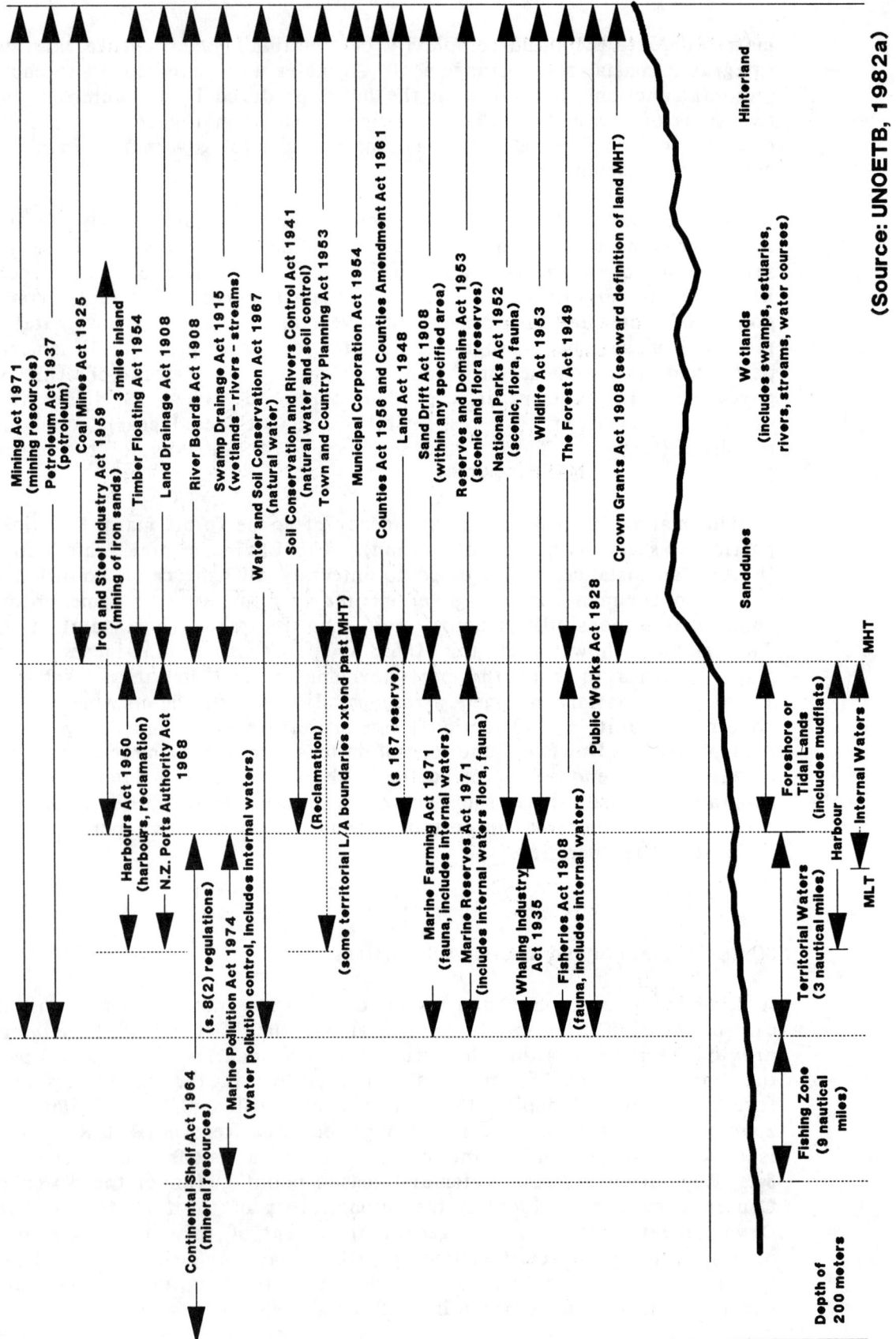

Mining Act 1971
(mining resources)

Petroleum Act 1937
(petroleum)

Coal Mines Act 1925

Iron and Steel Industry Act 1959
(mining of iron sands)

3 miles inland

Timber Floating Act 1954

Land Drainage Act 1908

River Boards Act 1908

Swamp Drainage Act 1915
(wetlands – rivers – streams)

Water and Soil Conservation Act 1967
(natural water)

Soil Conservation and Rivers Control Act 1941
(natural water and soil control)

Town and Country Planning Act 1953

Municipal Corporation Act 1954

Counties Act 1956 and Counties Amendment Act 1961

Land Act 1948

Sand Drift Act 1908
(within any specified area)

Reserves and Domains Act 1953
(scenic and flora reserves)

National Parks Act 1952
(scenic, flora, fauna)

Wildlife Act 1953

The Forest Act 1949

Crown Grants Act 1908 (seaward definition of land MHT)

Harbours Act 1950
(harbours, reclamation)

N.Z. Ports Authority Act
1968

(s. 8(2) regulations)

Marine Pollution Act 1974
(water pollution control, includes internal waters)

(Reclamation)

(s 167 reserve)

(some territorial L/A boundaries extend past MHT)

Marine Farming Act 1971
(fauna, includes internal waters)

Marine Reserves Act 1971
(includes internal waters flora, fauna)

Whaling Industry
Act 1935

Fisheries Act 1908
(fauna, includes internal waters)

Public Works Act 1928

Continental Shelf Act 1964
(mineral resources)

Hinterland

Wetlands
(includes swamps, estuaries,
rivers, streams, water courses)

Sanddunes

MHT

Foreshore or
Tidal Lands
(includes mudflats)

Harbour

Internal Waters

MLT

Territorial Waters
(3 nautical miles)

Fishing Zone
(9 nautical
miles)

Depth of
200 meters

(Source: UNOETB, 1982a)

Figure 8.3: Longitudinal Division of the Coastal Zone for Osaka Bay, Japan

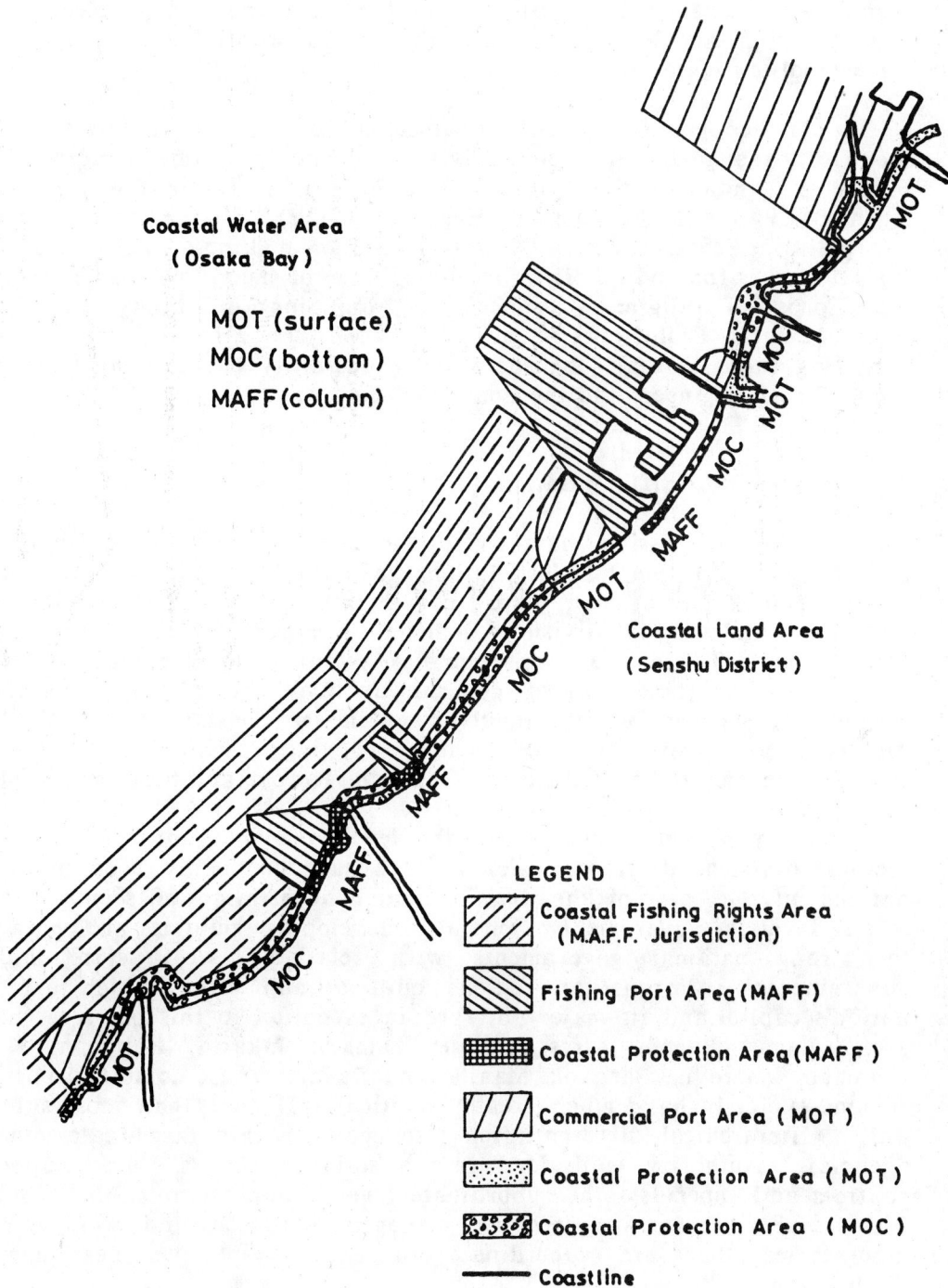

Coastal Water Area
(Osaka Bay)

MOT (surface)
MOC (bottom)
MAFF (column)

Coastal Land Area
(Senshu District)

LEGEND

Coastal Fishing Rights Area
(M.A.F.F. Jurisdiction)

Fishing Port Area (MAFF)

Coastal Protection Area (MAFF)

Commercial Port Area (MOT)

Coastal Protection Area (MOT)

Coastal Protection Area (MOC)

Coastline

Source: Osaka Prefecture Coastal Jurisdictions Map , Undated.

(Source: Shapiro, 1984b)

shoreline often is the jurisdictional boundary for national or state laws. Accordingly, government agency responsibility for the same sectoral function often changes at this boundary (see also Figure 3.1). Moreover, the geographic division of jurisdiction occurs lengthwise along the coast as well as across the coastal zone divisions. Figure 8.2 depicts New Zealand's cross shore divisions and Figure 8.3 illustrates a longshore division of coastal area in a metropolitan region of Japan.

To further complicate the institutional picture, different agencies are frequently assigned different activities within the same sectoral function. Figure 8.1 indicates that the Coast Guard (CG), Environmental Protection Agency (EPA), and the Corps of Engineers (COE) all have some responsibility for pollution control policy setting and plan making. The Coast Guard's jurisdiction is focused on oil spills and spillage of hazardous wastes from ships; the Corps of Engineers' concern is dumping of dredge spoil. The Environmental Protection Agency must consider all possible pollutants -- including those that are the specific responsibility of other agencies such as the Corps of Engineers and the Coast Guard.

8.1.3 Hierarchical differentiation.

The foregoing discussion characterizes the complexity in only a single-level of government involvement. Most nations have three or more hierarchical levels of government: national, state (provincial), and local (municipal). The functional and sectoral divisions depicted in Figure 8.1 can occur at any of these levels. To represent this complex picture, separate matrices could be constructed for each level of government involvement. An inventory and analysis of California state involvement in the coastal zones identified 42 different state units of government as having responsibilities that affect coastal zone uses and environments (Gamman, Towers, and Sorensen, 1974).

Generally, as one moves down the levels of government, the number of sectoral divisions decreases. Comparisons have commonly been made among nations on the basis of the division and concentration of sectoral functions among levels of government. Generally, developing countries are characterized by strong national governments with relatively weak state and local governments. Metropolitan regions built around cities which are both a nation's capitol and its major port are an exception to this pattern. Examples are: Lagos, Buenos Aires, Dar es Salaam, Jakarta, Rangoon, Kingston, Colombo, Monrovia, Bangkok, Manila, and Dakar. These combined port-capitol regions typically have much greater political influence than most subnational units. Hierarchical differentiation also occurs within large-scale government agencies. Authority is divided into a series of levels. Each upper level controls and supervises the subordinate levels. For example, the U.S. Marine Fisheries Service has both regional management offices and regional research laboratories that are subordinate to their respective headquarters in Washington.

8.2 Need for Sectoral Integration

The interconnection of important coastal-dependent economic sectors is the central reason for integrated coastal management in developing nations.

Integration of fisheries, tourism, oil and gas development, and coastal hazards regulation is essential because they share the same coastal zone as well as the same environmental and public service systems. For example, both fisheries and tourism depend on a high level of environmental quality, particularly clear coastal water. Both sectors receive spillover impacts such as pollution, loss of wildlife habitat and aesthetic degradation from uncontrolled economic development. Because fisheries require port services, while tourism depends on construction of an infrastructure system for water supply, sanitation, transportation, and telecommunications, they should be integrated with the transportation and public works sector.

Figure 8.4 indicates important linkages between coastal-dependent sectors. Positive and negative consequences of these linkages demonstrate the need for integrated coastal management in developing nations. For example, the need for coastal management is evident where coastal zone oil and gas development occurs in nations with a strong economic involvement in ports or fisheries. Indeed, a developing nation pursuing port development may be less likely to be concerned about integrated coastal management in the absence of a strong fisheries or tourism industry.

Coastal natural hazards are usually addressed in sectoral plans for public health and safety. These natural processes cut across all coastal-dependent economic sectors. Wind damage from a hurricane, inundation by a tsunami, or rapid coastal erosion can disrupt tourism, the fishing industry, port operations, public works, and transportation. Other sectors such as housing and industry are also vulnerable. The devastating consequences of development in coastal hazard-prone areas, together with the presence of any significant economic activity that depends on coastal resources or coastal location, necessitate integrated coastal management.

There is also strong evidence that the 30 or so developing nations in the humid tropics with extensive mangrove forests should have a strong incentive for integrated coastal resources management. IUCN's report, **Global Status of Mangrove Ecosystems**, documents that all developing nations with extensive mangrove forests are confronted with similar stresses which threaten the existence of this renewable resource (Saenger, Hegerl, and Davies, 1983). Conversion of mangroves to mariculture ponds or croplands is a particularly destructive and pervasive problem, in which three renewable resource uses are pitted against one another. The IUCN report recommends the preparation of national mangrove plans to protect and enhance this ecosystem's resource values. Such a nation-wide mangrove planning effort would represent a clear example of integrated coastal resources management.

In many small island nations, agriculture and forestry commonly occupy significant coastal upland areas where there is also strong pressure for conversion of these lands to tourism, vacation home estates, and, in some cases, housing for the resident population. Resolution of the conflicts arising from the conversion of forests or agricultural lands to housing or tourist facilities -- as well as the sedimentation impacts of forestry or agriculture practices on fishery habitats -- will require an integrated resource management approach.

Figure 8.4: Examples of Positive and Negative Relationships Among Sectors

SECTORS RECEIVING IMPACTS		SECTORS GENERATING POSITIVE AND NEGATIVE IMPACTS					
		Ports and Shipping	Transportation	Public Works	Fisheries	Tourism	Additional Sectors Can Be Portrayed
TOURISM	+	Passenger liner facilities	Airports and roads to tourist areas	Infrastructure for tourism development	Conservation of habitat areas vital to both recreational and commercial species		
	−	Ship pollution in beaches and swimming areas	Encroachment of urban dev. produced by roads	Encroachment of urban dev. produced by public works	Air & water pollution from fish processing industries & boats		
FISHERIES	+	Harbors and processing facilities	Roads and railroads to ship products	Sewage from processing		Sportfishing opportunities	
	−	• Ship pollution • Wetland fill	• Wetland fill • Estuary fragmentation	Wetland fill for development produced by provision of public works		• Pollution from fish processing • Fill of wetlands for tourist facilities	
PUBLIC HEALTH & SAFETY	+	Docks & channels enabling evacuation before storm & flooding	Roads & bridges for evacuation before storm & flooding	Public works such as dams & revetments to reduce or eliminate hazards	Fisheries development increases both amount & healthfulness of product	Improvement in quality of public water supply &./or sewage treatment	
	−	Port development in hazard-prone areas	Provision of transportation stimulates development of hazard-prone areas	Provision of public works stimulates development of hazard-prone areas	Water pollution from fish processing industries & boats	Development of tourist facilities in hazard-prone areas	
Additional Sectors Can Be Portrayed							

+ Sector reinforces or has positive impacts on another sector
− Sector has negative impacts on another sector.

8.3 Overview of Institutional Arrangements and Supplements

At least three major, permanent institutional arrangements are available for a nation-wide effort to broaden the scope of sectoral planning, and to create a structure for the resolution of coastal conflicts:

o **concentrate authority in a new centralized agency** (e.g. Cullen, 1982 (Australia); Amarasinghe and Wickremeratne, 1983 (Sri Lanka));

o **expand the duties of an existing agency** (e.g. Zile, 1982 (United States));

o **create a permanent interministerial council or network** to coordinate program management, policy making, land use allocation, or development (Zamora, 1979 (Philippines); Ecuador, Ministerio de Energia y Minas, 1988 (Ecuador)).

While adoption of these arrangements can advance the resolution of coastal disputes (Section 8.4), there may be many barriers to creation of a new, permanent institutional arrangement in some developing countries. First, key political leaders may not support a new arrangement for coastal governance. Second, strong sectoral agencies may express great reluctance at the potential loss of autonomy that might coincide with broadening participation and sharing of responsibility. Third, the economic costs of creating a new permanent bureau or council may be prohibitive. To respond to these difficulties, coastal managers can create an interministerial council as a supplement to a central agency (Amarasinghe and Wickremeratne, 1983 (Sri Lanka)). Such a council could supplement either a new agency or an expanded agency, and could be either temporary or permanent.

Still other supplemental arrangements are available to increase the responsiveness of sectoral agencies and expand participation in decision making. Such temporary strategies, usually invoked to set policy for specific issues or to resolve site-specific conflicts, include the following:

o **create an ad hoc panel** to guide policy or to organize a fact-finding process to clarify technical issues (IUCN, 1980; McNeely and Miller, 1983);

o **engage a facilitator** and convene a policy dialogue among key coastal actors to recommend specific policies or programs (McCreary, 1987);

o **engage a mediator** to lead parties through a face-to-face negotiation of specific conflicts (Susskind and McCreary, 1985; Susskind and Cruikshank, 1987);

o **engage a minister or other respected intervenor(s)** (other than the courts) to arbitrate a dispute (Klapp, 1984; Zhung, 1985);

o   finally, there is usually the option of seeking a
     legal remedy through the judicial system. However,
     the judicial system is unlikely to formulate a
     complex remedy such as a coastal zone management
     program.

While a nation would normally have only one major institutional
arrangement at a time, this would not necessarily preclude supplementary
strategies. For instance, a nation that depends on traditional sectoral planning
might supplement this approach with a dialogue among key agencies on critical
national issues, such as estuary pollution or coastal erosion. A nation that
creates a new agency might want to retain the option of organizing a mediated
negotiation process to settle a conflict between fishing and oil industries.

8.4 Elements of Choice in Organizing an Institutional Arrangement

Starting from the norm of fragmented, narrowly-focused sectoral planning and
development, it is important to stress that nations have a number of choices in
updating or augmenting their institutional arrangement. The **purposes** of
revising institutional arrangements, as was established above, are to integrate
development among sectors, anticipate and avoid negative impacts, establish
cooperative working relationships, share useful information, and create policies,
plans and projects capable of being implemented. In short, the goal is to
create an organizational climate that can help anticipate, avoid, or resolve
conflicts that dissipate the value of coastal resources and environments. Three
dimensions of choice in designing a revised institutional arrangement are the
degree of permanence, the **scope of participation**, and the **sharing of decision
making responsibility**.

8.4.1 Degree of permanence.

Table 8.2 illustrates a continuum of permanence ranging from a one-time
dialogue, to an interministerial commission to write a plan, to a permanent
agency to implement the plan. Of course, even the structure of a "permanent
agency" may change substantially over five to ten years.

8.4.2 Scope of participation.

The question of **who gets to participate** is vital in conflict resolution. Table
8.3 represents a continuum of possible levels of participation. At one extreme
is reliance on a single executive to render a decision. At the other end of
the spectrum is drawing on a broad set of representatives from private
industry, government industries, and non-governmental organizations. Choosing
an appropriate level of participation requires choices and tradeoffs. While the
quickest solution is to involve as few agencies or interest groups as possible,
this does not always produce an implementable or durable outcome. On the
other hand, a institutional arrangement that includes dozens of actors may
take years to reach a satisfactory outcome. Often a middle ground is arranged
to draw representatives from major agencies and key resource users.
Sometimes "tiers" of participation are an effective way to involve a broad

Table 8.2: Some Potential Elements of a National Institutional Arrangement for Coastal Management Arrayed on a Continuum of Permanence

Single Day Dialogue	Blue Ribbon Panel to Resolve Panel Specific Conflict	Interministerial Council to Formulate Broad National Coastal Policy	Permanent Agency
Days	Weeks	Months	Years
Less Permanent	-->		More Permanent

Table 8.3: A Continuum of Participation in Institutional Arrangements for Coastal Conflict Resolution

Single Executive	Single Agency	Multiple Agencies or Single Agency + Key Resource Users	Multiple Agencies + Resource Users	Multiple Agencies + Industry + Subsistence Users + NGOs
Less Participation	-->			More Participation

group while relying on a core group of negotiators to accomplish most of the work.

8.4.3 Sharing of responsibility.

As used here, "conflict resolution" includes a range of applications. One element of coastal conflict resolution is identification and avoidance of disagreements over broad policies for coastal resource allocation. At a slightly narrower degree of specificity, conflict resolution could be focused on resolving debates over environmental pollution standards or allocation of land to different types of uses. A still finer focus might involve a conflict over the details on the use of a specific site.

Participants in coastal conflict resolution might have a low, moderate, or high degree of responsibility in the formulation of coastal policy. The lowest level of responsibility would entail contributing information to an agency that does not have adequate resources or authority for information collection and analysis. The information contribution should enable the recipient to broaden the scope of its sectoral planning. A slightly higher level of responsibility would involve giving secondary ministries or bureaus the opportunity to review and comment on specific projects. A third level of sharing of responsibility would be a stake in reviewing and commenting on policy for specific issues, which would in turn guide decisions on a number of projects. The highest level would involve sharing of responsibility for policy making or project approval.

Table 8.4 illustrates a variety of possible institutional arrangements and a range of responsibility. The "+"'s indicate probable responsibilities for each institutional arrangement. The figure suggests that a moderate degree of sharing of responsibility will be accomplished by the permanent institutional arrangements. For instance, an existing agency might be willing to broaden the planning process and offer opportunities to review and comment on projects. Temporary supplements shown in Table 8.4 are likely to involve greater sharing of responsibility.

8.5   Major Institutional Alternatives to Broaden the Scope of Sectoral Planning

8.5.1 Create a new centralized agency.

Creation of a new agency can help set the stage for resolution of coastal conflicts by bringing a new, broad perspective to integrated coastal zone management. Such an agency could pinpoint prospective conflicts between development of important coastal-dependent sectors. A new centralized agency might also serve as a clearinghouse for efforts to assess and avoid harmful environmental impacts of coastal area development. Although a new agency could be beneficial, it should be understood that such a move represents perhaps the highest level of effort and commitment of funds of the strategies discussed. Sri Lanka has adopted this approach, and a careful analysis of that country's experience will hold valuable lessons for other nations.

Under legislation enacted in 1981, Sri Lanka created the Coastal Conservation Department (CCD), a problem-oriented agency that focuses its

Table 8.4: Sharing of Responsibility for Coastal Policy Making in Permanent Institutional Arrangements and Temporary Supplements

Degree of Sharing of Decision Making Responsibility Among Actors

LOW --> HIGH

	Broaden Planning Process	Project Review & Comment	Policy Making Review & Comment	Policy Making or Project Approval
PERMANENT INSTITUTIONAL ARRANGEMENTS				
Existing Agency	+	+		
Existing Agency with Advisory Council	+	+		
Expanded Agency	+	+		
Expanded Agency with Advisory Council	+	+		
New Agency		+	+	+
New Agency with Advisory Council		+	+	+
Interministerial Council of Equals		+	+	+
TEMPORARY SUPPLEMENTS				
Ad Hoc Panel		+	+	+
Facilitated Policy Dialogue		NA	+	+
Mediated Negotiation		NA	+	+
Arbitration		NA	+	+

attention on the solution of specific coastal conflicts and coordinates its work with other agencies responsible for guiding coastal development. The CCD has a three-fold mission modeled closely after several state programs in the United States. These include preparation of an inventory of the coastal zone, preparation of a national Coastal Zone Management Plan, and regulation of specific development activities through a permit process. The legislation empowers the Director of the CCD to call for environmental impact assessments of proposed development projects and provides for development review.

Another provision modeled after the U.S. is that the CCD reviews projects proposed by other agencies as well as by private developers. The CCD tries to avoid conflicts with sectoral agencies by encouraging early meetings with agency proponents of coastal development projects. It works with agencies to "scope" necessary environmental assessments. This linkage was designed to avoid difficulties that would arise if action were delayed until a formal permit application was forwarded to the CCD.

Close linkages were also created between the CCD and the Urban Development Authority (UDA), which has jurisdiction in a one kilometer coastal belt around the entire periphery of Sri Lanka. Development applicants must gain the approval of both agencies. The agencies, in turn, seek reciprocal advice, and each contributes to the training of staff.

Specific policies of the Act are designed to prevent or avoid use conflicts. Proposed development activities are not to cause any of the following impacts:

o   discharge unacceptable levels of effluents;

o   reduce the quality of beaches;

o   dislocate fishing activities;

o   affect the ecosystem adjacent to a marine sanctuary;

o   preempt a wildlife reserve.

A recent review of coastal management in Sri Lanka provides a comprehensive perspective on the program's history from its inception in 1981 to 1988 (Lowry and Wickremeratne, 1989). In the years since 1981 the Coast Conservation Department (CCD):

has conducted a significant amount of research and has prepared a Master Plan for Coast Erosion and a Coastal Zone Management Plan. It also issued 764 permits [between 1983 and 1987] for development activities, organized seminars and workshops on several aspects of coastal management, and developed effective relationships with agencies which have management responsibilities in coastal areas.

In the United States, the State of California created the State Coastal Conservancy to help resolve coastal use problems which could not be resolved through regulation. A companion to the strong regulatory Coastal Commission, the Conservancy is empowered to help plan and fund projects that create public access to the coast (Neuwirth and Furney-Howe, 1983), rebuild urban waterfronts (Grennell, 1988; Petrillo, 1988), preserve agricultural land, consolidate existing platted lots into orderly patterns of new development, and enhance streams and wetland environments (McCreary and Zentner, 1983; McCreary and Robin, 1985). Although the Conservancy is a state agency, it virtually never "imposes a solution." Its staff works closely with affected interests and peer agencies involved in coastal management. Often the Conservancy intervenes in a quasi-mediating role to try to find solutions agreeable to private developers, conservation interests, and local and state government (Susskind and McCreary, 1985). The agency has the authority and capacity to retain technical experts as needed, and may undertake grants to local governments, special districts, and nonprofit organizations.

The Conservancy is still unique in the United States, although other states are expressing interest in adapting some of the Conservancy's techniques. The major obstacles for transfer of this model among states appear to be (1) gaining political acceptance for a public agency charged with "beneficial development"; (2) resistance on the part of other existing agencies; and (3) obtaining the necessary funds.

8.5.2 Expand the powers and duties of an existing agency.

Another possible institutional arrangement is to expand the powers of an existing agency charged with some aspect of coastal resources management. Possible advantages associated with expanding the powers and duties of existing agencies are that the agency and staff bring a storehouse of expertise and experience and a network of contact, and would be thus be more readily accepted than a brand new agency. This approach has been used in the Philippines, where the National Environmental Policy Council (NEPC) is housed within the Ministry of Human Settlements. The NEPC was the lead agency in preparing a Coastal Zone Management Plan, using existing inventories of coastal resources, issues, and institutions. However, the NEPC has fallen short in gaining supportive legislation or a solid implementation structure for its plan (UNOETB, 1985).

In 1988 a summary of coastal management issues, objectives, and options for Ecuador concluded that:

> There are already in place sufficient laws and authorities to properly manage coastal resources. New laws are not necessary What is required is better coordination and enforcement of existing legislation (Ecuador, Ministerio de Energia y Minas, 1988).

Accordingly, the implementation of a national coastal program called more for a strengthening of existing institutions than the creation of new institutions. The proposed institutional arrangement has the following three elements:

o   An Interministerial Committee for the Management
     of Coastal Resources to provide high-level
     government support to the program, to assure
     central, political and administrative backing to solve

103

conflicts, to obtain international support for the program, and to promote interagency cooperation.

o   Formation of a Ranger Corps for each of the Navy's seven Port Captain jurisdictions. The Corps will be comprised of enforcement personnel from all of the responsible coastal regulatory agencies in order to improve enforcement of existing statues throughout the coastal area.

o   Formal designation of special management zones by Presidential decree. Within these zones it is especially essential to improve coordination among government entities to deal with real or impending conflicts. An advisory committee would be created for each special management zone. The committees are to consist of citizens representing the full range of interests in the area, as well as representatives of appropriate local and regional government entities.

Costa Rica decided not to create any new governance arrangements for administering the various aspects of its shorelands restriction law. The 1977 legislation vested the Institute of Costa Rican Tourism with primary responsibility for developing and implementing a management program for the marine terrestrial zone (Chaverri, 1989).

8.5.3 Create a permanent interministerial council.

Creation of a permanent interministerial council for coastal resources can provide a structure to develop mutually beneficial working relations and exchange views on pressing coastal issues. Other tasks that might be taken up by such forums include discussion of broad policy goals and review of specific development projects, review of proposed budgets, and identification of impacts of projects or programs. Interministerial councils can also devote attention to identification of gaps in information or expertise and opportunities for joint support of applied research. All of these accomplishments can create the foundation for joint problem solving, thereby helping sectoral agencies to avoid conflicts and to realize joint gains as they work towards their respective missions.

Several developing nations have convened interministerial councils to accomplish broader scope sectoral planning or to guide regional plans for coastal areas. Often this action responds to the realization that no single agency has sufficient information, professional capacity, political support, or authority to undertake integrated planning for coastal resources. Such councils are usually convened by the national executive or his or her delegate. The success of such arrangements may depend on the administrative skill and political support of the convenor, and the willingness of the convenor to allow real give and take.

Indonesia and the Philippines have each organized interministerial councils to formulate coastal resource policy (Koesoebiono, Collier, and Burbridge, 1982;

Kinsey and Sondheimer, 1984). In Thailand, the prime minister convened an interministerial committee to guide the Eastern Seaboard Development program (UNEP, 1985). The Thai group was organized into five subcommittees, each headed by a minister. They addressed industrial development, ports, education, investment, and institutional development.

Some Latin American nations have also organized interministerial councils both to broaden the scope of sectoral plans and to coordinate sectoral planning and implementation efforts. In Ecuador, the Directorate of Maritime Affairs of the Ecuadorian Navy (DIGEIM) played a leading role in putting together national institutions at an international seminar for the purpose of discussing Ecuador's coastal problems and alternative solutions (Vallejo, 1987). And as previously mentioned, an interministerial committee is one component of the governance arrangement to implement Ecuador's coastal zone management program. In Brazil, the Navy convened an interministerial commission for research. CIRM evolved into a national maritime commission on security and international relations (Knecht et al., 1984). In Colombia, a presidential decree created the Colombian Commission on Oceanography, a coordination group of marine-related agencies to recommend ocean research. In other nations, the Law of the Sea negotiations have motivated agencies with coastal responsibilities to coordinate their activities. For example, Cameroon convened the National Commission on the Law of the Sea (UNOETB, 1985).

Although little evaluative information on interministerial councils is available, it appears that they have enjoyed different degrees of success. Brazil's CIRM recommended that research on maritime and coastal issues receive $50,000,000 in funding. The requested funding was subsequently allocated. In contrast, the work of the Colombian Oceanographic Commission has suffered because it lacked a close linkage to policy making or to allocation of funds (Knecht et al., 1984).

Convening an interministerial committee has the overall appearance of "doing something", but it does not necessarily produce results. The most frequent accomplishments of interministerial councils have been to collect information, prepare inventories, and write guidelines. Some agencies may participate passively in interministerial councils just to "go along for the ride." Some participants may hold the view that the problem, as defined, is not part of the agency's mission. Close documentation of the stated agendas for these groups, and their degrees of success or failure, should benefit the practice of integrated coastal zone management.

8.6   Supplements to Major Institutional Arrangements

8.6.1   Create an interministerial council as a supplement to a central agency.

Another option is to create an interministerial council to provide advice to a central agency and to represent the interests of important economic sectors and major coastal actors. Interagency councils can foster discussion across coastal sectors and among professionals of different disciplines. Interagency councils organized as supplements to central agencies can enable the application of relevant experience without increasing the staff of the lead agency.

In Sri Lanka, a 13-member council, the Coast Conservation Advisory Council, was organized to assist the work of the Coast Conservation Department. Members of the Council include representatives from the following ministries: Fisheries, Tourism, Shipping, Local Government, Home Affairs, and Industry. Other members include representatives of the National Aquatic Resources Administration (NARA), the Office of Land Commissioner, the Urban Development Authority, and the Director of Irrigation. There are also three non-government members, one representing the universities, one representing voluntary organizations concerned with the coastal environment, and one representing the fishing industry. So far, staff of the Coast Conservation Department judge this institutional arrangement to be working well (Wickremeratne, personal communication). This hybrid arrangement offers the advantages of both centralized authority and well-defined responsibility, as well as close interministerial advice-giving and consultation.

8.6.2 Create an ad hoc panel.

Creation of an ad hoc panel to investigate specific issues can be a useful supplement to central agencies or interministerial councils. Conflicts over coastal resources may involve disagreements over technical issues such as standing stocks of fish, the capacity of inshore waters to assimilate pollutants, or erosion rates of a barrier beach. Often these competing scientific claims can be addressed through a process of fact-finding by an ad hoc panel of experts and other key interests. The objective of such a process is typically to produce some consensus on or at least a greater degree of understanding of the scientific issues. Some of these processes stop short of offering a policy prescription, while others are carried into policy recommendations.

Participation in ad hoc panels is sometimes limited to scientific experts. In their ultimate form, ad hoc panels can be organized as processes of "joint fact-finding," with face-to-face interaction of scientists, decision makers, and interest groups (Susskind and McCreary, 1985; Susskind and Cruikshank, 1987). Some objectives of joint fact-finding are mutual efforts to define issues, identify relevant information, explain cause-and-effect relationships, and design a program of research or analysis. Organizing such an effort may involve substantial work to recruit credible scientific expertise.

Joint fact-finding offers decision makers an opportunity to ask questions of scientific specialists and to understand the assumptions about methods and data that underlie the conclusions they offer. Such processes also foster direct discussion between scientists and other interested parties. Joint fact-finding may be a preferable alternative to the practice of pitting opposing experts against each other to bolster the positions of one party at the expense of another. Such "adversary science" is often the norm in scientific disputes, but it may actually obscure the real technical issues rather than narrowing the range of scientific disagreement. Many developing countries have had experience in organizing ad hoc panels for cooperative fact-finding or other policy making purposes. Joint fact-finding may be a logical extension of this experience.

Many developing countries have experience with scientific collaboration designed to produce a common research agenda and protocols for sharing results of scientific findings. Some observers believe that the most significant progress in international environmental cooperation appears to have been made through multinational scientific investigations (Caldwell, 1985).

In Ecuador, an ad hoc group was convened to help scope out the work program for the development of an integrated coastal management strategy. The ad hoc panel is chaired by a designate of the Ministry of Energy and the Environment. Its members include the Subsecretary of Fisheries, the Subsecretary of Forestry, and a representative of the Navy. Other members include representatives of CONADE, ESPOL, the University of Esmeraldes, the fisheries sector, the mariculture sector, and private conservation organizations. The panel was expected to have a lifespan of about three years -- the period required for initial program development (Olsen, 1987).

Another example of an ad hoc panel is the work of the International Oceans Institute (IOI) and U.N. agencies to summarize the "state of marine pollution in the world's oceans." Interests affiliated with the FAO and the Group of Experts on Scientific Aspects of Marine Pollution (GESAMP) proposed the idea in 1969. A 1972 action plan urged GESAMP to "assemble scientific data and provide advice on scientific aspects of marine pollution using an interdisciplinary approach." A leading scientist at Scripps Institution of Oceanography coordinated the first report, **The Health of the Oceans** (Goldberg, 1976). The first update was published by UNEP in 1982.

A 1982 IUCN meeting featured more direct interaction between scientists and decision makers, as well as a more definitive policy outcome. The World Congress on National Parks included several workshops to "provide managers of protected areas in aquatic habitats with improved management principles" (Salm and Clark, 1985). At the meeting, scientists and managers reached consensus on a framework for protected areas, as well as future research needs (McNeely and Miller, 1983).

The collaborative work of IOI and IUCN might be a useful model for more extensive joint fact-finding for coastal resources within a given nation. Experience with these and other joint fact-finding exercises deserve attention as conflict resolution techniques. Raiffa (1983) gives an example of fact-finding under the aegis of a blue ribbon panel. He was asked by the Mexico City Department of Public Works to evaluate two competing airport siting proposals for Mexico City. The Ministry of Public Works, the Secretaria de Obras Publicas (SOP), favored construction of a new airport, while the Secretaria de Comunicaciones y Transportes (SCT) advocated modernizing the existing airport in Mexico City. Both SOP and SCT had adopted projections about how the two plans would address concerns such as airport capacity, land costs, safety, noise, and effects on the military. Each set of projections, of course, supported the respective agencies' preferred alternative.

Although the analysts were asked to do impartial decision analysis of the two alternatives, they proposed an alternative procedure to the prevailing atmosphere of adversarial relations. They suggested that the president, Luis Echeverria Alvarez, appoint an impartial panel to supervise the analysis and to structure debate in a joint problem solving atmosphere. Raiffa and Ralph Keeny concentrated on the decision President Echeverria confronted during his six year term of office. They argued that since future events will influence airport planning, it was wise not to adopt a future master plan. The SOP was persuaded to support a compromise proposal that would upgrade the existing airport and only partly commit the government to building a new airport.

The analysts were able to defuse the controversy in part because they were neutral third parties, capable of performing a thorough decision analysis. But they also succeeded by introducing the idea that future decisions can be made contingently.

8.6.3 Convene a facilitated policy dialogue.

Policy dialogues are carefully structured discussions (see Appendix F) on broad topics (**not** specific disputes) among traditional adversaries or parties whose interests are suspected to be in conflict (Susskind and McCreary, 1985; Susskind and Cruikshank, 1987). Before a specific dispute arises over allocation of a resource or setting an environmental quality standard, a third party helper can help work through disagreements. Trained facilitators can sometimes intervene to create a forum for collaborative problem solving. Such a forum is often termed a facilitated policy dialogue. The roles filled by a facilitator include helping to structure an agenda, guiding discussion in an orderly fashion, and perhaps recording major ideas put forward by members of the group.

Policy dialogues can function as a research tool "for forecasting the politically viable middle ground that will ultimately emerge" (Gusman, 1981). An important question in structuring a policy dialogue is whether it must be limited to describing areas of agreement, or whether it will provide a forum for negotiating compromise positions. According to Gusman, if the agency to ultimately receive the findings has a habit of ignoring study panel recommendations, then participants would have little incentive to compromise.

If the participants in dialogue have adopted the rule of consensus, then each party owns a veto power. Under these circumstances, a ministry might choose to respect the dialogue participants as the action group to prepare public policy. Similarly, the participants would have an incentive to negotiate earnestly if they believe they will exert strong influence on public policy. If the group formulates a policy not in the ministry's interest, then it can effectively veto the outcome (Gusman 1981).

A facilitated dialogue may well be applicable to the early stages of policy formulation in developing nations. A study by the OAS Department of Regional Development (DRD) offered these conclusions:

> . . . environmental issues must be dealt with as early as possible during planning to avoid unnecessary conflict in the development process. "Environmental impacts" arising out of development are frequently conflicts between different resource users. Identifying these potential conflicts early on and exploring alternative development solutions to minimize or avoid the conflicts are therefore important goals of DRD regional development studies . . . Resolution of conflicts was far easier to negotiate during Phase I when parties were "equals" than it would have been if these conflicts have been discovered later after investments of

time, funds, and prestige had been made . . . Also, at this stage, positions of local interests have not hardened (Rodgers, 1984).

In Indonesia the task of establishing a setback in mangroves was accomplished through a series of discussions among several agencies (Turner, 1985). A policy dialogue process could have been very helpful in arriving at the setback. A facilitated negotiation could also be used to help identify the most pressing coastal issues, choose appropriate management strategies, or explore the terms of a critical area designation.

In the United States, facilitated dialogue and negotiation has been especially successful in the early stages of policy formulation (Susskind and McMahon, 1985). Agencies responsible for enforcing environmental law have been active participants. The U.S. Environmental Protection Agency has successfully organized two exercises in negotiated rule making, and is planning a third. EPA's standard procedure for arriving at environmental regulations is to depend on the judgments of agency staff, then publish the draft rule in the U.S. Federal Register, whereupon comments are invited. Often, proponents of weaker or stronger rules sue the EPA and costly litigation ensues.

Several experiments in negotiated rule making have been organized in the United States to avoid the litigation that often accompanies setting environmental standards. Representatives of key interest groups concerned with specific regulations engaged in facilitated discussion over the precise details of rules. In the rule making cases undertaken to date, the negotiating group convened a one or two day meeting each month. The total negotiation lasted from five to nine months, though longer or shorter time frames are equally probable. Industry, environmental groups, and state and federal agencies are all represented. Subcommittees of the full negotiating group may be formed as needed to delve into specific issues and report back to the full group. A "resource pool" is made available to help defray costs of attendance and to underwrite the costs of retaining technical experts to help clarify the issues at hand (Schneider and Tohn, 1985). Each negotiator has an equal voice, and each has a veto since the objective is to secure a consensus from all parties. Interviews of the participants in the EPA's first two negotiated rule making efforts indicated a high degree of support for the process (Susskind and McMahon, 1985).

8.6.4 Organize a mediated negotiation process.

If communication has broken down and ministries or resource users are frozen into antagonistic positions, a neutral outsider acceptable to all sides can help reopen discussions by serving as a go-between. With appropriate technical knowledge, mediators can help disputants invent ingenious solutions to problems that make joint gains possible for all parties (Susskind and Cruikshank, 1987).

Mediators typically take a more active role than facilitators. While they help to set agendas, run meetings, and record minutes, they may also meet individually with key actors in a dispute to better understand the interests that underlie the positions each actor takes in the negotiation. Mediators may propose packages of their own after hearing the interests of all sides. A mediator may also assist individual negotiators in presenting the outcome of a mediated effort to their constituency or agency.

This has been the recent experience in site planning controversies over Southern California wetlands and conflicts between fishing and oil interests in Santa Barbara. In a case involving local fishing interests and several oil companies, a mediator helped the parties work through disagreements over geophysical testing and the location of vessel traffic lanes. One of the outcomes was essentially an exclusion zone, which fishing interests agreed not to enter at certain times (Susskind and McCreary, 1987).

Facilitated policy dialogues and mediated negotiation have each been used in several coastal conflicts. Appendix F delineates the steps involved in using these processes.

8.6.5 Arbitration.

In some cases, when the normal administrative machinery fails to produce an acceptable outcome, the intervention of a respected nonpartisan party (or parties) may be useful. Binding arbitration represents a higher degree of intervention than facilitation or mediation, since an arbitrator generally has the authority to impose a solution. In Indonesia, the Minister of Finance resolved a conflict between the state-owned oil corporation (Pertamina), other industrial sectors, and the International Monetary Fund. The solution involved restructuring Pertamina and cutting back the company's operations (Klapp, 1984).

In China, plans are underway to create an arbitration commission to resolve coastal management disputes, as part of the implementation of the **Law of Coastal Zone Management of the People's Republic of China**. According to a leading coastal planner based in Nanjing:

> The commission is to arbitrate the disputes arising between the boundaries of the provinces and counties, and between the developing industries. The members of the commission should be experts from all spheres of life. They should be of special knowledge, justice, and high prestige among the masses. The commission will carry on its work in a certain democratic procedure. The arbitration has special superiority of its own. It is superior to the judgement in the court. Because the arbitration commission [can] carry out the principles of equality, democracy and consultation, its decision is easily accepted by the two parties of the dispute. That will be of benefit to the solution of the dispute thoroughly (Zhung, 1985).

Although the details of the Chinese Proposal are sketchy, it is noteworthy that what appears to be a standing arbitration service is under consideration.

8.6.6 Seek a legal remedy through the judicial system.

If the normal administrative machinery fails to produce an acceptable outcome, major coastal actors may have the option to seek a legal remedy in the judicial system. Supplementary institutional arrangements, such as policy dialogues and mediated negotiation, are often represented as alternatives to the judicial system. However, in most nations, the judicial system remains the last resort. In complex disputes over resource allocation the judicial system is not likely to have access to the necessary technical information to render a wise and well-informed decision. Another drawback of the judicial system is that it normally deals with disputes with only two parties, while many coastal resource conflicts involve multiple parties.

NOTE: Our literature search has not produced any reports of coastal management disputes in which two other approaches to dispute resolution-- mini-trials and non-binding arbitration -- have been used, but they deserve to be mentioned to complete the list of available techniques.

9. PROGRAM EVALUATION

Coastal managers need good information about the success or failure of management strategies and governance arrangements. Program evaluation serves this goal. Implementation and evaluation are distinct, yet related steps in the overall process of coastal management. Implementation is "the delivery of specific objectives set forth in constitutionally adopted public policies" (Mazmanian and Sabatier, 1983). Two conditions are required if program implementation is to be evaluated: (1) an adequate post-implementation time period to allow a program to reach maturity; and (2) a set of indicators for measuring performance.

All evaluation studies seek to assess program performance, although they differ markedly in the evaluative criteria employed. Two basic types of evaluation can be distinguished. One type of evaluation focuses on the policy making process (such as the number of permits issued), and the other type focuses on the eventual outcomes (such as improvement in public access to the coast). Of course, many evaluations measure both process and outcomes.

Process evaluation examines the means by which goals are achieved. Process indicators include the clarity of goal statements and legislative mandates, measures of the rationality of organizational structures and the process and information flow, the adequacy of yearly budget allocations, the number of permits issued, and the number of agreements executed to promote interagency cooperation.

Outcome evaluation measures the extent to which the program's goals or objectives are achieved. Outcome indicators can be subdivided into instrumental factors and environmental/socioeconomic conditions. Instrumental indicators measure goals whose achievement is thought necessary to the achievement of environmental and socioeconomic goals. These may include the extent of the information base, the efficiency of permit review, and the extent of public participation. Environmental or socioeconomic conditions measure such things as the extent of protected wildlife habitat or the number of jobs created. Table 9.1 presents a list of indicators that have been use to evaluate ICZM efforts.

One of the first evaluations made of a coastal zone management program used the **outcome** or goal approach to effectiveness assessment. This entails:

> discovering what the organization itself has postulated as its ideals, then . . . measuring organizational success by objective observation of the degree to which the standard is reached. An organization may thus be judged effective to the extent that it achieves its goals (Swanson, 1975).

A myriad of environmental and socioeconomic outcome indicators can be identified. Some environmental indicators are water quality, the amount of protein derived from coastal fisheries, linear kilometers of the coast in public ownership, and acreage of wetlands protected or restored. Socioeconomic indicators include lives and property lost due to coastal hazards, tonnage of

**Table 9.1: Examples of Process Indicators and Outcome Indicators
for Evaluating Coastal Zone Management Programs**

PROCESS INDICATORS

o   budget allocation per year
o   number of permits issued, denied, conditional
o   consistency of law dealing with coastal
       management
o   number of agreements or memoranda executed for
       interagency cooperation
o   availability of appropriately trained and
       educated staff
o   number of subnational programs initiated or
       approved
o   quality of information used in program
       development

OUTCOME INDICATORS

Instrumental factors

o   cost and length of time for permit review
o   number of procedures and steps eliminated
       ("streamlining")
o   public participation -- number of
       individuals and groups
o   geographic scope and issue coverage of
       information base

Environmental or socioeconomic conditions

o   water quality (dissolved oxygen, nutrient levels)
o   fishery yields
o   protein component of diet derived from coastal
       fisheries
o   number and linear distance of access ways
o   kilometers of coast in public ownership
o   number of recreation user days
o   number of coastal species on the International
       Union for the Conservation of Nature and Natural
       Resources (IUCN) endangered species list
o   acreage of wetlands protected or restricted
o   number of low and middle income housing
       units provided within the coastal zone
o   tonnage and value of commodities handled in ports
o   employment derived from fisheries, ports and
       tourism sectors
o   hazard impacts -- lives lost, property damaged

114

goods handled in ports, and employment generated by ports, fisheries, and coastal tourism.

9.1 Outcome and Process Assessments of Coastal Management Programs

For developed countries, particularly the United States, there is an emerging literature of coastal program evaluation. Coastal zone management programs in the United States at both the Federal and state levels have been a testing ground for many innovative and ambitious institutional arrangements and environmental management strategies. Passage of the Coastal Zone Management Act in 1972 was the first nation-wide program in land use management. The 18 year track record and innovation of numerous federal and state programs have induced scholars and environmental management practitioners to conduct numerous evaluations.

In the outcome evaluations, goals are measured by various indicators. Swanson's 1975 outcome evaluation described the effectiveness of the Bay Conservation and Development Commission (BCDC) in preventing land fills, increasing public access, and improving shoreline quality. McCrea and Feldman (1977) reviewed Washington state's first three years of experience with the Shoreline Management Act. The program was judged a success in minimizing environmental damage, enhancing public access, and encouraging water-dependent uses. Healy (1978) evaluated the impact of the first thirteen months of the 1972 California Coastal Zone Conservation Act on beach access and implementation, density and economic growth, wildlife habitat protection, energy facilities development, aesthetics, and agriculture.

Sorensen (1978) examined a series of instrumental indicators for nine U.S. states that have similar collaborative management arrangements between state and local governments. Under the collaborative arrangement, local governments are required to prepare land use plans based on state guidelines. The plans are then reviewed by the state coastal agency. The instrumental goals were: reduce uncertainty, develop an affirmative policy position, manage resources of state or regional concern, manage resources that extend beyond local government boundaries, accommodate local variation, and facilitate accountable decision making.

Rosenbaum (1979) reviewed enforcement of and compliance with coastal wetland regulations in several states including Massachusetts, New Jersey and North Carolina. McCrea (1980) completed an evaluation of output indicators for port planning for the State of Washington. She specifically examined conformance of port projects with the goals of the State Shoreline Management Act.

Process evaluations, by contrast, do not assess whether goals have been achieved, but rather whether the organizational structure and political process will facilitate those goals. Cullen (1977) has reviewed Australia's Port Phillip Authority with an emphasis on the intergovernmental conflicts that arose in a particular site. Sabatier (1977) reviewed permit procedures and policy directives of the 1972 California Coastal Zone Conservation Act analyzing a random sample of regional permit decisions appealed to the State Coastal Commission. Decisions were analyzed in terms of the major issues discussed, and the decisions reached by regional and state commissions on different types of development. The Conservation Foundation described the enactment of the 1972 California Coastal Initiative, its planning and permitting activities, and its re-enactment in the Coastal Act of 1976 (Healy, 1976).

The U.S. Office of Coastal Zone Management (OCZM) and its successor, the Office of Ocean and Coastal Resource Management (OCRM), are directed by Section 312 of the Coastal Zone Management Act to prepare annual evaluations of each state's coastal program. As with most government evaluation directives, these "312" evaluations concentrate on process -- a simpler evaluation than focusing on outcome. Measuring input is almost always easier than measuring output. The 312 program evaluations typically emphasize process indicators such as the number of programs approved and the funds allocated to different functions. The most recent Biennial Report to Congress for 1980 and 1981 (U.S. National Oceanic and Atmospheric Administration, 1982), for example, highlights interesting or innovative program features, but does not assess the degree to which the goals of the program have been achieved.

It would be extremely useful to do program evaluations that would compare nations or subnational units with similar coastal environments and different levels of mandated coastal management in order to assess the impact of coastal management programs on environmental and socioeconomic outputs. For example, a useful comparison would be Georgia, with no coastal management program, Florida, with a modest level of coastal management, and North Carolina, with a concerted coastal management program. All three states have similar coastal environments. So far, there are no assessments that compare the effects of varying levels of a program with the effects of no program. Consequently, policy makers may draw erroneous conclusions about the impact of a coastal management program.

There is also the problem of causation. It is often difficult to determine the extent to which improvements in coastal environmental indicators (such as a decrease in water pollution) can be attributed to one governmental program. For example, an analysis of federal responsibilities in state coastal programs by the U.S. Department of Commerce, Office of the Inspector General (1983), documents a tendency by OCRM to attribute all improvements in coastal environmental quality to programs administered by OCRM, even though many other agencies have programs that directly or indirectly improve coastal environmental and socioeconomic conditions. The EPA, for instance, may be the key actor in cleaning up water pollution, despite similar efforts of the national and state coastal management program.

Costa Rica and Sri Lanka are the only two developing countries that have ICZM programs with an implementation history. Costa Rica's program implementation dates from 1978 and Sri Lanka's implementation began in 1983. Accordingly these are the only two countries where evaluations have been conducted on program implementation. A case study of Costa Rica's program has just been completed and will be published in **Coastal Management** (Sorensen, 1990).

Lowry and Wickremeratne (1989) concluded in a recent publication that:

> a detailed examination of Sri Lanka's program suggests that its
> strength and vigor are due in large part to (1) strong coastal

orientation of the country; (2) widely shared agreement about what the coastal problems are, what the causes of the problems are, and to a lesser extent, what the appropriate roles of government are in dealing with the problems; (3) a law that provides a strong legal basis for management; (4) strong program leadership; (5) adequate political support for planning and management; and (6) an adaptive, incremental approach to the development of the planning and management program.

Their chapter in **Ocean Yearbook 7,** from which this quote is taken, provides partial support for this important finding. More extensive data and analysis will appear in a book on the history of the Sri Lanka program. The book is being prepared by the International Coastal Resources Management Project at the University of Rhode Island and Sri Lanka's Coast Conservation Department.

Few developing nations have evaluated either program process or outcome. Public administration in developing nations has not adopted the concept of program evaluation. Also, in nations where authority is highly centralized within the chief executive's office, program evaluation will not be practiced if the results could be negative or critical. If program evaluation is to be both an effective and efficient process, a "checks and balances" relationship must exist among the legislative, executive, and judicial branches of government. Since this authority relationship often does not exist within developing nations, program evaluation will probably occur in rare and unusual circumstances -- such as when the action or inaction of a government agency causes an unexpected disaster, examples of which would be catastrophic flooding or the sudden crash of an important fishery. Therefore, practitioners of coastal management evaluation in developing nations are unlikely to be within the national government agencies. They will come from the organizations providing international assistance for coastal resources management such as USAID, the World Bank, regional development banks, and IUCN. International assistance organizations should be motivated to conduct evaluations of coastal management they have supported for the same two basic purposes motivating all organizations that practice program evaluation: (1) to assess the efficiency and effectiveness of the investment; and (2) to determine what improvements should be made to the program.

9.2 Criteria to Assess Program Implementation

Two methods could be used to gather information for a process evaluation or implementation assessment. The first method is to review and synthesize existing case studies of coastal management programs in developed and developing nations. This is feasible for a mature public policy field where a large data base exists. In the absence of recorded case studies, new case studies could be undertaken. Clearly, such an effort is outside the scope of this report. The second method for determining the criteria for implementation is to review implementation analyses in other policy areas, extract the relevant portions, and organize a framework for assessing coastal management implementation.

We choose this second method given the sparse data on coastal management in developing countries and the uneven data for developed nations. The most useful analytic framework was constructed by Mazmanian and Sabatier (1983). Their framework is derived from a study of five widely divergent policy areas as well as the early experience of the California Coastal Commission.

Mazmanian and Sabatier list six preconditions for success:

o   mandates are clear and consistent;

o   mandates incorporate a sound theory identifying causal linkages to policy objectives; enabling act gives implementing officials sufficient jurisdiction over target areas and points of leverage;

o   enabling legislation structures implementation process to maximize the probability that implementing officials and target groups will perform as desired;

o   leaders of the implementing agency possess managerial and political skills and are committed to statutory goals;

o   the program is supported by organized constituency groups and a few key legislators;

o   priority of objectives is not undermined over time by the emergence of conflicting policies.

This list addresses all the organizational problems identified in Section 5.5 of this report. The authors are realistic about applying the framework:

> In practice, of course, all conditions are very unlikely to be attained during the initial implementation period for any program seeking substantial behavioral change . . . In short, the list of conditions can serve not only as a relatively brief checklist to account post hoc for program effectiveness or failure but also as a set of tasks which program proponents need to accomplish over time if statutory objectives are to be attained. In fact, the appropriate time span for implementation analysis is probably seven to ten years. This gives proponents sufficient time to correct deficiencies in the legal framework, and it also tests their ability to develop and maintain political support over a sufficient period of time to actually be able to bring about important behavioral or systematic changes. It also gives the political system sufficient experience with the program to decide if its goals were really worth pursuing and to work out conflicts between competing values (Mazmanian and Sabatier, 1983).

With these caveats, the Mazmanian and Sabatier framework is useful for organizing a discussion of coastal management implementation. Some analysts question the applicability of implementation evaluation criteria derived from studies in developed western societies to the developing world. In fact, we find that the criteria posed by Mazmanian and Sabatier are congruent with implementation principles derived from studies of developing nations (Esman,

1978; Soysa, Chia, and Collier, 1982). Also, three companion case studies to the first edition of this document (Towle, 1985; Turner, 1985; Hayes, 1985) support one or more of the criteria. The International Institute for Environment and Development's review of developing nations' environmental management programs concluded that:

> The most important prerequisites for effective operation seem to be the mandate, the organizational structure and the level of professional competence with which the institution is endowed. Where an institution's success has been marginal, or where it has failed, one can usually pinpoint the absence of one or more of these factors (International Institute for Environment and Development, 1981).

The difference in implementation assessment between developed and developing nations is not in the nature of the criteria but in the relative importance each plays in program achievement and failure. For example, the size and competence of professional staff is usually cited as one of the major implementation obstacles in developing nations. By comparison, developed nations usually do not rate staffing as a major problem. Conversely, developed nations are commonly beset by constituency support problems and opposition by target groups. These participatory process obstacles are seldom expressed as a concern by developing nations.

Many of the examples used to illustrate the seven conditions for successful implementation are drawn from California, either the San Francisco Bay Conservation and Development Commission (BCDC) or the state Coastal Commission. This focus on California reflects the more developed literature for that state. California's ambitious, innovative and controversial coastal management programs have made it the favorite case study for program evaluations. Also, the comparatively long history of BCDC (25 years) and the California Coastal Commission (17 years) present an opportunity to assess the evolution of implementation -- an opportunity not present in other coastal programs.

9.3 Clear and Consistent Policy Objectives

Rhetoric often clouds policy objectives in many environmental management programs. This problem is evident in both the Federal Coastal Zone Management Act and several state acts in the United States. Because mandates are the result of negotiated legislative compromise, they often avoid making clear statements of priorities among apparently conflicting goals. In a few cases, a vague mandate may be sharpened by rules and procedures adopted subsequent to legislative authorization.

Swanson's review of the Bay Conservation and Development Commission (1975) showed that much of the agency's success could be traced to its specific mission. Clear rules were laid out for making decisions on three issues: bay fill, public access, and improving the visual quality of the shoreline. He concludes that "BCDC owes much of its success to working for clearly defined goals through the political process."

119

In developing nations, vague and conflicting goals appear to be a common problem in environmental management programs. The goals and objectives emerge from the accumulation of laws over the years. Because outmoded legislation has often been kept on the books, administrators must choose among an array of vague and conflicting mandates (International Institute for Environment and Development, 1981).

9.4 Good Theory and Information

This criterion argues for the availability of good information about the consequences and opportunities for coastal development. Impact assessment is well understood as a task of tracing impacts from uses and activities through to biological/physical changes and social consequences. For instance, filling shallow mudflats around a bay margin is likely to have a negative impact on fisheries. Constructing buildings in a way that impairs public access will mean diminished recreational opportunities, aesthetic appreciation, and possibly lost tourism revenues. While the causal networks of environmental impacts such as those outlined in Chapter 5 may be well understood, there is far less understanding about the ecological function of a natural system or the analytic tools needed to predict magnitudes of impact with certainty.

A major weakness of the Regional Seas program is the lack of data and models to make persuasive cause and effect connections between the terrestrial pollution sources and the degradation of the marine environment (Hulm, 1983). Several analysts have commented on the importance and the difficulty of achieving a sound technical data base for coastal management.

Clark (1978) assessed the recruitment of natural science expertise in the preparation of the California Coastal Plan. He concluded that while scientists who had participated in various phases of plan preparation felt that the scientific content was adequate, there were many areas where the planner-scientist linkage could be strengthened.

During the course of his work, Clark interviewed a number of coastal management specialists on the theoretical and informational basis for the plan. One analyst observed:

> This data-rich situation is mandatory to the success of the planning program they've just adopted; otherwise it will be just a political judgement (Dickert quoted in Clark, 1978).

Another analyst suggested that lack of good technical information weakened the Commission's ability to defend its jurisdictional boundary:

> Because the commission did not have good information about the dynamics of coastal systems -- the [coastal zone border] was often set too close to the coast, and in some cases too far back . . . Since the Coastal Commission does not have the scientific information to support the line for much of its length, it appears that the inland boundary in several locations will be moved coastward during this session of the legislature, further compromising the Commission's ability to develop and implement a management strategy . . . (Clark, 1978).

McCrea and Feldman (1977) noted that implementation of the Washington Shoreline Management Act was hindered during its first few years by lack of information on natural systems upon which local government could make permit decisions. The same situation probably existed in all state coastal programs during the first phase of implementation.

McCreary (1979) identified three series of barriers to the use of biological information in California's coastal zone planning. Some are generic to the practice of biological research, a second set is common to any effort to inject scientific information into environmental planning, and a third set is specific to California's coastal planning. Despite the apparent chasm between scientific information and the planning process in California, Mazmanian and Sabatier (1983) gave the coastal program high marks in comparison to the other social programs reviewed.

In the absence of good natural resource information, the tendency is to base resource allocation decisions almost entirely on economic or political considerations. Analytic tools for quantifying values in monetary terms are relatively well developed. Economic indicators (dollars invested, jobs created) may have more meaning than measures of wildlife habitat acreage, catch per unit effort, species diversity indices, or rates of runoff and sedimentation. The U.S. experience strongly suggests that to be successful in coastal zone management, programs in developing countries must incorporate a strong knowledge of coastal processes along with the economic calculus. The information should be expressed in concise policy terms. When there is a potential for two policies to conflict, such as developing marinas versus protecting shellfish nurseries, clear decision rules should be laid down for reaching an outcome (California deals with this by prescribing very narrow conditions under which wetlands can be filled).

Reports on environmental management programs in developing nations have consistently noted the lack of adequate data and maps for environmental assessment and policy making (USAID, 1979). Since coastal management programs require an information rich base, we expect that data and map limitations will be expected to be a major problem in program development and implementation.

9.5 Sufficient Jurisdiction and Authority

Implementing officials should have sufficient jurisdiction over target groups and other points of leverage to attain the coastal program's objectives. The discussion in Chapter 5 identified insufficient planning and regulatory authority as a major problem. To ensure that an agency has adequate authority, at least three important decisions must be made about:

o   the geographic scope of the jurisdiction;

o   the types of projects and issues within the agency's jurisdiction;

o   the functional responsibilities (e.g., permit letting, advisory, review and comments, capital allocation, etc.) assigned to the agency.

The respective geographic jurisdictions of the San Francisco Bay Commission (BCDC) and the California Coastal Commission present an interesting contrast. BCDC's jurisdiction extends just 100 yards inland from the Bay margin. It excludes wetlands diked off from tidal action. BCDC's narrow jurisdiction is conducive to a crisp focus on the issues of preventing bay fill, preserving access, and reviewing designs for bayside structures. In these areas, it has a strong, proven record (Swanson, 1975).

This same narrow mandate has several shortcomings. Davoren (1982) points out that the agency's pioneering effort to promote access (and block bay fill) "never grew beyond the 1969 land use law and, more significantly, the one agency that the public sees as controlling the Bay's destiny does not have power over undeveloped shoreline use or Bay waters to exercise that control." The California Coastal Commission, on the other hand, has an inland jurisdiction which varies in width from 200 feet landward of mean high tide in urban areas, to five miles around important estuaries and wetlands. Depending on the section of coast, this has given the Commission authority to intervene in a wide range of issues that affect the quality of the coast -- watershed erosion, urban coastal design, traffic capacity on roads near the shore, and offshore drilling in state and federal waters. With these broad powers, the Commission's attention is diffused by being drawn in many directions at once. Unlike BCDC, the Coastal Commission cannot concentrate on a few issues in a narrow geographic jurisdiction.

Functional responsibilities may be concentrated in a single agency or divided among several agencies. Where responsibilities are divided, the lead agency for coastal resource management must gain the cooperation of other agencies -- a process that may prove slow and difficult, as suggested by the experience of both developed and developing countries.

A new agency for the coast will likely have to contend with a wide array of existing institutions and agencies, both formal and informal. There may be battles over jurisdictional "turf," budget, staff, political influence, and organized constituencies. An analysis of the relationship between the California Coastal Commission and state agencies pointed out that the new agency:

> joined one of the nation's largest and most active state bureaucracies, a collection of commissions, boards and agencies noted for both professionalism and fierce independence (Banta, 1978).

Banta found that the Commission did not assign a high priority to resolving conflict with other state agencies. Conflicts continually arose when the Commission exercised its authority to review plans and permits of sister agencies and when the Commission's tight deadlines necessitated making decisions without formal consultation with the experts of other agencies in fields such as water quality and fisheries biology. When the 1976 California Coastal Act was drafted creating a successor agency, steps were taken to improve interagency coordination (Banta, 1978). Similarly, a review of Costa Rica's administrative alternatives for a coastal resources management program (Blair, 1979) suggested that a new agency would be unwieldy, and would have difficulty gaining cooperation from other agencies for sound resource management.

In Sri Lanka, the approach of building an interagency network for coastal management has been quite successful. The Coast Conservation Department (CCD), a unit within the Ministry of Fisheries, has established effective working relationships with three key agencies -- the Urban Development Authority (UDA), the Tourist Board and the Central Environment Authority (CEA). The four agencies share a commitment to incorporating environmental and coastal zone considerations into their decision making, and are mutually supportive. CCD has a permit authority for a zone 300 meters landward of the mean water mark, UDA has permit authority over all development activities within one kilometer of the shoreline, and CEA is an umbrella agency responsible for formulating environmental assessments and setting pollution control standards. Until the CCD gained an independent permit authority, the Tourist Board referred proposals for development of 50 units or more to the CCD for guidance on necessary beach setbacks (Kinsey and Sondheimer, 1984).

Neither a broad jurisdiction nor a narrow jurisdiction is "right" -- both have advantages and disadvantages. A developing nation must choose whether to concentrate on a narrow jurisdiction at the risk of overlooking important coastal issues, or gaining broad jurisdiction with the responsibility of resolving numerous issues and perhaps dissipating its energies in numerous directions.

9.6 Good Implementation Structure

The implementation process should be structured to maximize the probability that implementing officials and target groups perform in a manner to attain the objectives of the coastal program. The key ingredients for a successful coastal resource management program are an adequate budget, a sympathetic and dedicated host agency, and adequate political support. According to McCrea and Feldman (1977), the commitment of the responsible state agencies was critical to the successful achievement of the Washington State program's goals.

Agencies that are created as autonomous units who can control their own staffing and budget process have a greater likelihood of achieving program objectives. An example is Sri Lanka's creation of the Coast Conservation Department to administer the nation's coastal zone management program. It was elevated to departmental level to provide the agency with budgetary and administrative flexibility (Kinsey and Sondheimer, 1984).

By contrast, the Philippines has a weak implementation structure. The National Environmental Protection Council (NEPC) has established a coastal zone management program that addresses such comprehensive issues as ports and dredging, tourism development, and marine pollution. There is also an Interagency Coastal Zone Task Force with representatives from 22 different agencies of government. However, there seems to be little attention paid by other agencies to either the Task Force or NEPC's coastal management program. The critical elements of cohesive policy regarding coastal resources and coordinated implementation among agencies to achieve these policies has not been articulated (Kinsey and Sondheimer, 1984).

The state of Florida had a similar administrative problem to the Philippines. The state had a very small coastal management agency. Authority for coastal permitting and resource management was disbursed among several agencies.

The initial program activity in Florida, preparation of a coastal atlas, was not initially linked to regulation or resource management and the implementation process was not clear.

"Target groups" for coastal management are numerous. The groups regulated by U.S. coastal management are primarily land developers, industrialists, commercial recreation entrepreneurs, and others seeking to extract or exploit resources. Relevant groups may also include the fishing community, which would benefit from the control of water pollution but which might perceive catch restrictions as a disadvantage to them. The tourist industry would benefit from provision of beach access, control of water pollution, and preservation of natural phenomena that attract visitors. Local environmental interests represent a fourth type of target group.

Much of the early success or failure of coastal management programs in the United States is due to the involvement of target groups. The early evidence from developing countries also suggests that participation of target groups or constituencies is crucial to program success. The initial legislative mandate creating the BCDC was successful in large part because a citizens' organization mounted an effective public relations and lobbying campaign (Odell, 1972). They were, in turn, encouraged and reinforced by the work of BCDC's technical staff (Swanson, 1975).

Legislative support was as crucial to California's success with coastal management as support from conservation interest groups. The momentum generated in the early days of the Commission carried through to strong legislative support for the 1976 Act to create a successor agency and program. The persistence of people who cared about coastal conservation was credited as a key factor in the passage of the 1976 Act (Duddleson, 1978). Final negotiations included some successful bargaining with advocates of the building trades, thus overcoming some of the major opposition to the legislation.

Maine's early experience with integrated coastal resource management, in contrast to California's, was colored by the failure of state planners to create a constituency of support among local citizens and communities (Lewis, 1975). Even though several progressive environmental laws were already on the books when the program was launched in 1972, the proposed state level intervention did not have broad-based citizen support and opposition interests killed the program. Six years later, the program was resurrected by the state with a greater measure of citizen involvement, expanded local government involvement, and greater emphasis on encouraging desirable economic development such as shellfish harvesting.

Creation of a CRM program in American Samoa depended on involving target groups in each step of program development and thereby building a supportive constituency (Templet, 1986). Both instrumental goals (government process) and outcome goals (resource protection and economic development) emerged from a series of meetings with village councils. The next step was to distill a list of specific management policies from the general goals. Existing agencies with responsibility for some aspect of coastal management were encouraged to participate in return for increased staff and outside technical assistance.

American Samoa had several statutes which, when combined, provided sufficient authority to implement a coastal area management program. The program was invoked by executive order based on existing legislation. Consistent with traditional Samoan orientation to graphic presentation, the first atlas of American Samoa was produced. This was both a valuable analytic device and an educational tool complemented by courses on coastal resources in the public schools. Consensus decision making, cooperation among agencies and broad education were the hallmarks of American Samoa's coastal program. The Samoan program benefitted from western techniques, but retained the essence of traditional culture.

Turner (1985) and Towle (1985) have suggested that involvement of residents of coastal areas is a strong precondition for successful coastal program implementation. Siddall, Atchue, and Murray (1985) have cited the participation of a shrimp producers' association as a potentially important factor in successful mangrove management.

9.7 Staff Competence and Commitment

Successful coastal management demands both executive skill and strong staff level capabilities. Coastal issues span a wide variety of disciplines: marine and terrestrial biology, hydrology and hydraulics, engineering, site planning, architecture, policy analysis, and economics. Often the skills of specialists in these diverse fields must be brought together to bear on a single coastal project. In developing countries, skills and expertise are often in short supply so training programs are a crucial necessity. For example, Indonesian coastal managers cited the lack of trained and experienced personnel in many fields related to CRM including management/administration, policy analysis, data gathering and research, and enforcement (Kinsey and Sondheimer, 1984). International assistance efforts to promote coastal management clearly need to provide training in a series of disciplines and in techniques to bridge disciplines to structure a team approach.

Leaders of the implementation agency should possess substantial managerial skill and be committed to achieving the program's objectives. Without strong, politically adept leadership, it is doubtful that a program for integrated coastal management can pass its infancy, let alone grow into maturity. An excellent example of the dividends paid by skilled political staff is the California coastal program. In its first incarnation after the 1972 initiative, the California Coastal Zone Conservation Commission drew some of its key staff, notably its executive director and the chairman of the Commission, from the BCDC. These individuals had perhaps the most experience possible given the short history of coastal zone management at the time. After the first Commission's Plan was translated into legislation by the 1976 California Coastal Act, a new complement of top staff arrived who had participated in drawing up the mandate for the new agency. They were not just legal technicians, but were politically seasoned by their experience with the first coastal agency and the negotiations to secure passage of the 1976 Act. This legislative experience and network of contacts has well served the Coastal Commission and a companion agency, the California State Coastal Conservancy, in securing funding and retaining jurisdiction.

9.8 Maintaining the Program's Priority on the Public Agenda

This condition for effective implementation can be compromised by a large number of factors -- many beyond the control of an agency. Most of the state coastal programs in the United States have had to withstand an economic recession and two "energy crises" with attendant pressure for accelerated offshore oil and gas exploitation. Dramatic shifts in a national economic policy or a change in administration usually bring changes in priorities for government intervention. These pressures underscore the importance of maintaining a strong organized constituency for coastal management, as described in subsections 9.4 and 9.6. In developing countries, coastal management may depend on ensuring that sustained-yield of mangroves or coastal fisheries stays near the top of the public policy agenda.

One strategy for keeping coastal management at the top of the public agenda in a developing nation may be to forge strong liaisons with other sectors of the government or national economy. Templet's review of the program in American Samoa tends to confirm this suggestion (Templet, 1986).

NOTE: Application of the evaluation framework to the Costa Rica program. During 1988 and 1989 an evaluation was conducted of Costa Rica's shoreland restriction program (Sorensen, 1990). The evaluation started with a tour of both coasts, reading the literature that pertained to the program, and interviewing 23 persons involved with the process. Next, analyses were done for each of the program's components. Process and outcome indicators were then identified and measured. The final section of the evaluation applied the eight criteria just described. The criteria worked well. They explained both why the Costa Rica program was achieving its objectives and why it had encountered implementation problems.

10. RECOMMENDATIONS

Our recommendations are divided into two parts. The first part suggests programmatic recommendations for international assistance organizations. The second offers more specific guidelines for national coastal resources managers and administrators.

10.1 Programmatic Recommendations for International Assistance Organizations

1. **Communicate the availability of a broad array of coastal resource management strategies.**

   International assistance organizations should foster awareness that a broad array of institutional arrangements and a broad array of management strategies are available to help developing nations manage coastal resources and resolve coastal use conflicts.

2. **Fund an array of coastal programs, using a variety of institutional arrangements and management strategies.**

   No single prescription can be applied uniformly to every developing country. Management strategies must be tailored to reflect the institutions, laws, and customs now in place. Strategies must also reflect the geographic extent and severity of issues, and the available expertise and staffing.

   International assistance organizations should support programs that represent each type of institutional arrangement. Such an investment strategy would recognize that there is no "best" institutional arrangement for managing coastal resources. "Goodness" in an institutional arrangement can best be judged by the effective and efficient resolution of coastal use conflicts. In the near future, coastal resources and environments in most developing nations will be managed by fragmented sectoral planning and development programs. Reversing resource degradation and improving management practices (such as achieving sustained-yield in these nations) may only occur with international assistance that is structured to work through the existing sectoral planning programs.

   International support to broaden the scope of sectoral programs -- such as port development, offshore oil development, fisheries development, and natural areas protection -- may yield more environmental improvement than integrated planning in the short-term. Similarly, assistance in non-integrative management strategies such as critical area designation, shoreland exclusion zones, and environmental impact assessment may pay immediate, low-cost dividends. Advice, information, and financial support to accomplish these goals are urgently needed.

127

3. **Demonstrate how integrated coastal management is linked to improvement in socioeconomic conditions in order to build support for improved and sustainable resources development.**

Coastal zone management programs are usually initiated as a response to a perceived use conflict. Launching a coastal program demands a clear motivation. Usually an event that dramatizes the importance and vulnerability of coastal resources is needed. A severe decline in a resource or a devastating experience with natural hazards may trigger new initiatives. Over the long-term, the socioeconomic benefits of coastal resources management must be evident in order for environmental quality and natural area protection to enjoy continued support. **Maintaining fisheries productivity, increasing tourism revenues, sustaining mangrove forestry, and avoiding the costs associated with natural hazard devastation are four compelling arguments for integrated coastal resources management.**

In less developed large islands or continental nations without fisheries, mangrove forestry, tourism, or natural hazard devastation as important national concerns, there may be little potential for the initiation and implementation of an integrated coastal management program. An infusion of considerable funds and expertise from international assistance organizations may be needed. Among these nations, the management strategies of environmental impact assessment, shoreland exclusion zones, or critical area designation might be excellent strategies to improve management of coastal resources and environments.

4. **Fund research to identify the major obstacles and aids to successful implementation of integrated coastal resources management in developing countries in order to inform effective design of future programs.**

Research is needed to clearly identify the obstacles that block effective coastal zone management. The set of conditions needed to ensure program success needs to be verified and refined. In the U.S., successful program implementation depends upon clarity of goals, understanding of cause and effect relationships, and a strong constituency. Analyses of environmental programs in the developing world suggest that vaguely worded goals and lack of expertise are two serious obstacles to program implementation. Non-governmental organizations, such as NATMANCOMS, the Eastern Caribbean Natural Areas Management Program (ECNAMP), and IUCN appear to be the most important expression of constituency influence on coastal resources management in the developing world. Research is needed to determine which conditions for successful implementation suggested by the U.S. experience should be emphasized in coastal management programs in developing nations.

5. **Compile a global issues index to assist international assistance organizations in setting program priorities and organizing integrated coastal management projects, and to promote international information exchange.**

The international practice of international coastal zone management can be advanced by understanding the finite number of ways that the world's development activities can have an impact on the world's coastal resources and environments. These impacts can be described in terms of cause and effect linkages. These linkages can, in turn, provide the basis for a global listing of coastal issues. A structured survey should be conducted to compile a national roster for each issue. The survey should also rate the issues according to their importance to each coastal nation and to international organizations. The format presented in Appendix B is proposed as a framework for the global issues index. Nations identified in connection with each issue could form the basis of international networks for sharing information and expertise. (The IUCN document **Global Status of Mangrove Ecosystems** (Saenger, Hegerl, and Davie, 1983) represents a good model for a compilation of issues related to one type of renewable resource.)

6. **Devise a common format to analyze a nation's existing institutional arrangements in order to guide the creation of programs for integrated coastal resources management.**

Every coastal nation has established its own institutional arrangements -- laws, customs, management strategies and organizations -- to allocate coastal resources. A common analytic framework is needed to clearly reveal the institutional complexity of coastal management. Such a framework should reveal structural and functional divisions, geographic and activity subdivisions, and levels of government.

Over the last decade, many descriptions of national and state approaches have been generated. Many reports presenting such descriptions are cited in this book and in articles in journals such as **Coastal Management** and **Ocean and Shoreline Management**, and conference papers have produced an important body of literature. To promote more useful comparative assessments, the editors of professional journals, as well as convenors of future conferences, should work to establish a consistent format to describe institutional arrangements. The format should be compatible with USAID's environmental profile series in order to facilitate the incorporation of coastal governance as a component in forthcoming reports.

7. **Organize forums for evaluation and exchange of experience on the application of specific management strategies.**

This report identified 11 management strategies and several institutional arrangements which have been used to help resolve coastal conflicts in developing nations. Resource managers in developing nations should benefit by direct communication with their colleagues to explore and reflect on their experiences in formulating and applying the different management strategies and institutional arrangements. Such a dialogue already exists among international practitioners of environmental impact assessment, and could provide a useful prototype and building block.

8. **Support the development of national technical capacity for coastal zone management.**

Coastal nations need a minimum information base, a set of analytic methods, and a cadre of professionally trained staff to conduct coastal area management. One element of such an information base is a set of maps or an inventory capable of graphically representing coastal environments, resources, and hazards. Another element of the information base should consist of a review of existing conflicts and a forecast of future coastal use conflicts. The review should identify the causes and consequences of impacts and the major actors involved. A review of existing and potential conflicts should provide guidance in setting national priorities for future coastal research and data collection.

9. **Use consensual styles of decision making such as policy dialogues and mediated negotiation as supplemental arrangements for conflict resolution.**

Consensual styles of decision making such as policy dialogues and mediated negotiation may hold promise as supplementary institutional arrangements for conflict resolution. Policy dialogues, facilitated by a nonpartisan intervenor, can be useful in scoping the major issues to be addressed in coastal management and in helping to clarify areas of scientific disagreement. Face-to-face communication among scientists, policy makers, and other major coastal actors have been very helpful in policy making and conflict resolution. Specific case studies are needed to test and document the application of these supplemental techniques in the developing world. A careful effort is also needed to forecast major constraints and opportunities for application of these dispute resolution techniques.

10. **Train coastal resource managers in negotiation and conflict resolution techniques.**

Successful policy making requires both the application of scientific talent and wise methods of dealing with the major actors in coastal management. Many coastal managers have received specific training in environmental science and regional planning. International assistance organizations are providing additional training in cartography and other skills. To enhance the chances for implementation of policies and management strategies, managers should build their capabilities in negotiation and bargaining. Assistance organizations should supplement their programs with training in negotiation and dispute resolution. Training programs must be matched to the cultural, institutional, political and economic conditions of the respective developing nations.

10.2   **Guidelines for National Coastal Zone Managers and Administrators**

1.   **Tailor boundaries for integrated coastal management programs to "capture" and enable resolution of the relevant coastal issues. Simple political jurisdictions or rigid "zones" may be ineffective in promoting successful integrated coastal resources management.**

Coastal managers and administrators can choose from a broad array of possible coastal zone boundaries (Figure 2.1). Some boundaries are quite narrow and are best suited to deal with use conflicts occurring at the immediate shoreline. If watershed-generated impacts pose use conflicts, then a coastal program boundary extending inland to the ridge line of watersheds draining into the coast would be more appropriate. The seaward side of the boundary should be adjusted to reflect the economic significance of the fisheries and ports sectors and the importance of inshore spawning and rearing habitats.

2.   **The key characteristics of coastal nations, together with important coastal issues, should guide the choice of coastal management strategies for a nation.**

Several factors must be considered in designing management strategies for a nation. These include the economic importance of coastal-dependent sectors, the extent of prior governmental experience with some aspect of coastal resources management, experience with the destructive consequences of coastal hazards and the revenue available for program implementation.

3.   **Encourage broad participation in development and implementation of coastal programs and coastal zone management. Artisanal resource users, the scientific community, and NGOs should be recruited as participants, along with representatives of key government bureaus with a stake in coastal management.**

Coastal zone management requires broad participation from groups outside government. NGOs, in particular, can represent key coastal users, communicate government policies, assist in training resource managers, and compile relevant information. In this way NGOs can stretch the capacity of government agencies, provide realistic insights into possibilities for effective program implementation, and broaden the dissemination of important information.

4.   **Spell out the causal relationship between policy goals or rules and the protection and management of coastal resources. This is likely to improve the chances for successful implementation of an integrated coastal zone management program.**

Government programs with ill-focussed goals are difficult for lay people to understand and hard to administer. Such concepts as ecological cycles, food webs, and impact networks may not be well understood by lay people, so the logic of regulatory or planning strategies is often not apparent. Both the initial mandate and the implementation process should stress the links between resource management goals and the management strategy. Clearly explaining the rationale for rules and regulations can help build support for a management program.

5. **Coastal managers and administrators in all nations can benefit from applying the strategy of environmental impact assessment (EIA) to projects affecting the renewable coastal resources.**

EIA can offer a flexible procedure to identify, evaluate, and help avoid the worst impacts of coastal development. This strategy can be a part of a larger integrated coastal resources management program or it can stand alone. The technical quality of EIA can be strengthened through the incorporation of a coastal atlas or data bank as a source of information for the assessment.

6. **Resource managers from small island nations with important tourism and fisheries sectors are likely to benefit by developing and implementing programs for integrated coastal management.**

Almost every activity on small island nations is linked to coastal resources and environments. These linkages strongly suggest that management strategies be adopted to identify and resolve conflicts arising from competing demands on a small resource base.

Again, the precise strategies used to integrate planning and management across economic sectors should be tailored to local needs, capabilities, and traditions. Strategies to accomplish integrated management could include broad-scope sectoral planning, special area/regional planning, or national land use planning. Preparation of an island-wide resource atlas covering watershed lands, shorelands, and submerged resources is likely to be a valuable investment in protecting critical biological and economic resources. Small islands may require external expertise. The ECNAMP experience in preparing coastal atlases for 25 island areas in the Eastern Caribbean could provide a model because it combined imported skills with indigenous expertise. Middle income nations -- both large islands and continental nations -- with a strong fisheries or tourism sector, or a recent experience with the effects of coastal hazards, are also strong candidates for integrated coastal resources management.

7. **Consider using management strategies such as shoreland exclusion zones and critical area designations as an appropriate first step towards integrated coastal management.**

Nations can cope with problems that occur in a limited geographic area, such as shoreline erosion, loss of coral reefs, estuarine spawning grounds, or encroachment on endangered species habitats using the strategies of critical area designation or shoreland exclusion. Both strategies can be implemented on a site specific basis, commensurate with available information, staffing, and expertise. They can be reinforced with special area planning or broad-scope sectoral planning of a larger geographic scope. Combining estuarine or marine protected areas with more land-based strategies offers the possibility of managing an entire coastal ecosystem.

Since critical area designations and shoreland exclusion zones are relatively inexpensive and simple to administer, these strategies can be especially appropriate for nations that are otherwise without a strong coastal orientation.

8. **Consider launching integrated coastal management programs on a regional basis. If this approach succeeds, consider expanding it to national land use planning.**

The regional level of focus allows resource planners to concentrate on the most severe problems. The regional focus also enables a nation to obtain experience with integrated coastal management, provides time to develop and recruit expertise, and presents an opportunity to make needed midcourse corrections. In this respect, the coastal programs of developing nations could follow the same evolutionary process documented in a number of developed coastal nations.

In most developing nations, the coastal zone consists of several types of environments. The majority of the coastal zone consists of agriculture, rural settlements, pristine environments, or other undeveloped land. The remaining coastal zone usually consists of an urbanizing region surrounding the nation's major port and its associated estuary. Major port and estuary complexes are usually the locus of the greatest intensity and number of coastal resource conflicts, and the greatest environmental degradation. As a result, national interest in integrated coastal management has usually focussed on the need for managing regions defined by the metropolitan port and estuary complex.

Coastal management at the regional scale provides the opportunity to test new concepts and approaches as a pilot effort before committing energy and political capital to a nation-wide effort. The risk and consequences of a failure are likely to be considerably less when a program is implemented on a smaller scale. The experience gained during the regional effort should increase the likelihood of success of a later nation-wide effort.

**DATA NEEDED TO ASSESS THE VALUE OF A NATION'S
COASTAL RESOURCES**

Sector	Data-Needs
Coastal Fisheries	o Linear kilometers of coastline or square kilometers of coastal zone known to function as nurseries for finfish and shellfish
	o Number of harbors for fishing fleets
	o Number of existing mariculture facilities
	o Number of potential sites for mariculture
	o Estimated stock of commercial fin- and shell-fisheries that are biologically-dependent upon the nation's coastal zone
	o Catch (in tons) of commercial finfish and shellfish that are biologically-dependent upon the nation's coastal zone
	o Dollar value of total catch
	o Dollar value of internal consumption
	o Dollar value of export harvest
	o Tax revenues generated by fisheries
	o Relative contribution of fisheries to total GNP
	o Number of fish-processing plants
	o Dollar value added by processing plants
	o Number of nationals employed directly or indirectly by fisheries sector
	o Relative proportion of nationals employed as a function of the total workforce
	o Relative contribution of fisheries as a fraction of total worker earnings
	o Commitment to development of fishery sector indicated by (a) creation of a ministry; (b) legislative mandate or executive order; (c) preparation of sectoral plans; and (d) capital investment

Appendix A

NOTE: Data on the value of coastal fisheries is difficult to collect. First, the coastal-dependency of a species may not be well studied. Second, species have transboundary habits so it is difficult to attribute a standing stock to a single nation. Finally, the possibility of foreign ownership of some portion of the fishing fleet or fish-processing facilities complicates assessment of the actual contribution of the fishery sector to a national economy.

Coastal Tourism

o   Number of linear miles of coast allocated to coastal tourism development

o   Presence of swimmable beaches with excellent offshore water quality

o   Presence of coral reefs, bird rookeries, reserves, sanctuaries, and other wildlife-oriented areas

o   Extent of public relations effort for coastal tourism

o   Number of facilities built within 1000 meters of the coast

o   Infrastructure devoted to coastal tourist development

o   Dollars earned by coastal tourist-serving development

o   Tax revenues derived from coastal tourism

o   Relative contribution of coastal tourist facilities to GNP

o   Number of nationals employed directly or indirectly by coastal tourism sector

o   Relative proportion of nationals employed as a fraction of the total workforce

o   Relative contribution of tourism as a fraction of total worker earnings

Ports

o   Number of major and minor ports (as defined by the Ocean Yearbook)

o   Tonnage of imports and exports

o   Forecasted future exports

o   Number of ship and boat building facilities

o   Number of support facilities (e.g., chandleries)

	o	Size of port hinterland served
	o	Dollar value of exports
	o	Dollar value of imports
	o	Tax revenues generated by ports
	o	Relative contribution of ports to GNP
	o	Number of nationals employed directly or indirectly by port sector
	o	Relative proportion of nationals employed as a fraction of the total workforce
	o	Relative contribution of port sector as a fraction of total worker earnings
Hazards	o	Geographic extent of hazard-prone areas
	o	Frequency of major disastrous events
	o	Frequency of events causing major damage to lives or property
	o	Number of lives lost
	o	Number of injuries
	o	Number of structures damaged
	o	Dollar costs of reconstruction and relocation
	o	Dollar costs of service disruptions
	o	Insurance rate increases as a function of hazards
	o	Type and extent of architectural/engineering standards for development in hazard-prone areas
	o	Type and extent of standards for siting structures in hazard-prone areas
	o	Number of structures and dollar value built in hazard-prone areas
	o	Amount of vacant/uncommitted land available in hazard-prone areas
	o	Amount of vacant/uncommitted land available in non-hazard-prone areas

o   Commitment to intervention in hazard sector indicated by (a) creation of a ministry; (b) legislative mandate or executive order; (c) preparation of hazard guidelines for siting new development; and (d) preparation of architectural/engineering standards for development in hazard-prone areas

APPENDIX B

GLOBAL ISSUES INDEX

This appendix presents a preliminary global list of important coastal resources issues. Three types of issues are included: impact issues, hazards, and sectoral planning concerns. Under each specific category of issues, we list nations where the issue has been documented to occur.

Part I of this Appendix presents a set of causal networks. Each one flows from left to right. The sequence of events begins with the use of a coastal resource, which involves human activities. These activities produce changes in environmental or socioeconomic conditions, which in turn result in an impact of social concern. For simplicity, we have compressed the cause and effect sequence into three steps. This compression is achieved by combining uses and activities. Also, environmental and socio-economic condition changes often progress through several cause and effect sequences before culminating as an impact. For example, increased turbidity reduces light penetration, which in turn decreases or kills coral growth. This produces the impact of decreased yields from coral reef fishery stocks. We have compressed sequences of multiple condition changes into one step. Several are cross-referenced.

Parts II and III of this Appendix present hazards and sectoral planning concerns, respectively, and Part IV contains additional comments.

USE OR ACTIVITY	------------>	ENVIRONMENTAL CHANGE	------------>	IMPACT OF SOCIAL CONCERN

I. IMPACT ISSUES

A. Estuary, harbor and inshore water quality impacts.

1. domestic and industrial sewage and waste disposal	estuary pollution, particularly adjacent to urban areas	decreased fish yields

(Australia, Bangladesh, Barbados, Brazil (Rio and Sao Paulo), China, Cuba, Dominican Republic, Ecuador (Guayaquil), Greece, Guyana, Haiti, Indonesia, Israel, Jamaica, Japan, Kenya, Korea, Mexico, Morocco, Mozambique, Nigeria, Pakistan (Karachi), Panama, Philippines, Senegal (Dakar), Spain, United Kingdom, most states in the U.S., Venezuela)

2. domestic and industrial sewage disposal	estuary pollution	contamination of fish shellfish and water contact areas

(Ecuador, Japan)

USE OR ACTIVITY ------------▶	ENVIRONMENTAL CHANGE ------------▶	IMPACT OF SOCIAL CONCERN
3. tourism sewage disposal	estuary pollution	decreased fish yields
(Fiji, Jamaica)		
4. domestic and/or tourism sewage disposal	estuary and beach pollution	decreased tourism and recreation attraction
(Barbados, Jamaica, Trinidad (Tobago))		
5. flood control and/ or agricultural development, impoundments or diversions of coastal rivers	increased estuary salinity, decreased estuary circulation	decreased fish yields
(Australia, Bangladesh, India, Kenya, Senegal, Somalia, Sri Lanka, Tanzania)		
6. coastal oil development, chronic release of oil and/or large oil spills from accidents	oil pollution of estuarine and inshore waters	decreased fish yields, tainted fish and shellfish, decreased recreation or tourism quality
(Dominican Republic, Ecuador, Indonesia, Liberia, Mexico, Nigeria, Oman, Trinidad (Tobago), Venezuela)		
7. port development and shipping and/or off-shore shipping of oil, chronic release of oil and/or large oil spills from accidents	oil pollution of estuarine and inshore waters	decreased fish yields, decreased recreation or tourism quality
(Bangladesh, Barbados, France, Indonesia, Ivory Coast, Jamaica, Liberia, Madagascar, Morocco, Mozambique, Nigeria, Oman, Pakistan, Senegal (Dakar), Singapore, Thailand)		

USE OR ACTIVITY	------------>	ENVIRONMENTAL CHANGE	------------>	IMPACT OF SOCIAL CONCERN
8. agricultural pesticides		toxic pollution of estuaries and inshore waters		decreased fish yields, fish kills

(Bangladesh, Ecuador, Guyana, Mexico, Philippines, many U.S. states)

| 9. crop, grazing, mining or forestry practices in coastal watersheds | | watershed erosion, estuary sedimentation and increased turbidity | | decreased fish yields |

(Brazil, Bulgaria, Dominican Republic, Ethiopia, Haiti, India, Indonesia, Jamaica, Kenya, Madagascar, Malaysia, Mozambique, Philippines, Surinam, Tanzania)

| 10. crop, grazing, mining or forestry practices in coastal watersheds | | watershed erosion, floodplain deposition | | increased flood hazard |

(Brazil, Costa Rica, Dominican Republic, Haiti, India, Indonesia, Kenya, Malaysia, Madagascar, Mexico, Mozambique, Philippines, Surinam, Tanzania)

| 11. crop, grazing or forestry practices in coastal watersheds and inshore areas | | watershed erosion, increased sedimentation of bays, deltas, and port areas | | sedimentation of navigation channels and berths |

(Kenya)

| 12. crop, grazing or forestry practices in coastal watersheds | | watershed erosion, increased sedimentation, changed deposition of sediments in bays, deltas and inshore waters | | beaches covered with unattractive sediment, decreased recreation and tourism attraction |

(Kenya)

141

USE OR ACTIVITY ------------>	ENVIRONMENTAL CHANGE ------------>	IMPACT OF SOCIAL CONCERN
13. agricultural development and fertilizer	increased amount of nutrients entering estuaries, eutrophication pollution	decreased fish yields, fish kills
(Japan)		
14. coastal mining	increased sedimentation and turbidity, change in composition of bottom sediments	decreased fish yields
(Jamaica)		

B. Groundwater quality and quantity.

15. agricultural development	withdrawal of ground-water at rate greater than natural recharge, salt water intrusion of aquifer	contamination of groundwater for domestic and/or agricultural use

(India, Morocco, Mozambique, Oman, Thailand, Tunisia, United States (Florida, California, North Carolina), many coral atolls)

16. tourism and residential development	withdrawal of ground-water at rate greater than natural recharge, salt water intrusion of aquifer	contamination of groundwater for domestic and/or agricultural use

(Fiji, United States (Florida))

C. Filling of wetlands (including mangroves).

17. port development	filling of wetlands	decreased fish yields

(Australia, Fiji, Jamaica (Kingston), Japan, Korea, Nigeria (Lagos), Singapore, Western Samoa)

USE OR ACTIVITY	ENVIRONMENTAL CHANGE	IMPACT OF SOCIAL CONCERN
18. port development	filling of wetlands	decreased fishing or mariculture areas
(Japan)		
19. mining and soil disposal	filling of wetlands	decreased fish yields
(Fiji, Indonesia, Malaysia, Thailand)		
20. tourism development	filling of wetlands	decreased fish yields
(Fiji, Jamaica)		
21. residential development (particularly canal estates)	filling of wetlands	decreased fish yields
(Australia, Jamaica, Nigeria, United States (Florida))		

D. Mangrove impacts.

USE OR ACTIVITY	ENVIRONMENTAL CHANGE	IMPACT OF SOCIAL CONCERN
22. agricultural, maricultural or salt evaporation development	draining or diking of mangroves	decreased fish yields
(Australia (Queensland), Bangladesh, Ecuador, India, Indonesia, Panama, Philippines, Sri Lanka, Thailand)		
23. agricultural, maricultural or salt evaporation development	draining or diking of mangroves	reduction or loss of rare or endangered endangered species
(Dominican Republic)		

USE OR ACTIVITY ------------▶	ENVIRONMENTAL CHANGE ------------▶	IMPACT OF SOCIAL CONCERN
24. mangrove harvesting for wood chips, fuel-wood and building materials	harvesting at rate greater than sustainable yield, decreased productivity	decreased fish yields, decreased timber yield of successive harvests

(Bangladesh, Indonesia, Philippines, Thailand)

25. mangrove harvesting for wood chips, fuel-wood and building materials	harvesting at rate greater than sustainable yield, loss of habitat	reduction or loss of rare or endangered species

(Dominican Republic)

26. mining (usually tin)	local removal of mangrove forest	decreased fish yields

(Indonesia, Malaysia, Thailand)

E. Coral reef and atoll impacts.

27. municipal and/or industrial sewage disposal	coral reef pollution	decreased fish yields, decreased tourism and recreation attraction

(Jamaica, Kiribati, Philippines, Sri Lanka, United States (Hawaii))

28. coral mining	coral reef destruction	decreased fish yields, decreased tourism and recreation attraction

(Indonesia, Sri Lanka)

29. coral mining	coral reef destruction	increased shoreline erosion

(Indonesia, Sri Lanka)

USE OR ACTIVITY ------------▶	ENVIRONMENTAL CHANGE ------------▶	IMPACT OF SOCIAL CONCERN
30. coastal or offshore mining	sediment and turbidity, pollution of coral reefs	decreased fish yields, decreased tourism and recreation attraction

(Indonesia, Malaysia, Sri Lanka, Thailand, Western Samoa)

31. oil shipping along offshore international routes	oil pollution of offshore waters	decreased growth of coral reef, increased beach erosion, decreased tourism attraction

(Barbados, Jamaica)

32. dredging for construction materials	sediment and turbidity pollution of coral reefs	decreased fish yield, decreased tourism and recreation attraction

(Fiji, Western Samoa)

33. crop, grazing or forestry practices in coastal watersheds	watershed erosion, sediment and turbidity pollution of coral reefs	decreased fish yields, decreased tourism and recreation attraction

(Indonesia, Jamaica, Kenya, Philippines, Sri Lanka, United States (Hawaii))

34. fishing with dynamite	coral reef destruction	decreased fish yields, decreased tourism and recreation attraction

(Barbados, Tanzania)

35. intensive, localized fishing effort	harvesting at rate greater than sustainable yield	decreased coral reef associated fish yields

(America Samoa, Cook Islands, Tahiti, United States (Hawaii))

USE OR ACTIVITY ---------->	ENVIRONMENTAL CHANGE ---------->	IMPACT OF SOCIAL CONCERN

F. Beach, dune and delta impacts.

36. recreation and/or tourism development | trampling of beach and dune vegetation | initiation or increase of shoreline erosion, increased hazard

(Australia, United Kingdom)

37. recreation and/or tourism development | trampling of beach and dune vegetation | decreased tourism and recreation attraction

(Australia, Denmark, Germany, United Kingdom, most U.S. states)

38. grazing of livestock | trampling and/or overgrazing of beach and dune stabilizing vegetation | initiation or increase of dune migration onto agricultural areas or infrastructure

(Bangladesh, Kenya, Mozambique, Oman, Somalia)

39. mining beach sand | removal at rate greater than natural accretion | initiation or increase of beach shoreline erosion, increased hazard, loss of native vegetation, wildlife habitat and natural amenities, decreased tourism attraction

(Australia, Dominican Republic, Comoros, St. Lucia, Western Samoa)

40. flood control and/or agricultural development and impoundment or diversions of coastal rivers | decreased supply of beach material to shoreline | initiation or increase of shoreline erosion, increased hazard

(Bangladesh, Kenya, Mozambique, United States (California, Louisiana, Texas))

USE OR ACTIVITY	------------>	ENVIRONMENTAL CHANGE	------------>	IMPACT OF SOCIAL CONCERN

G. Fishing effort.

41. intensive and extensive fishing effort — harvesting at rate greater than sustainable yield — decreased fish yields

(Greece, Italy, Jamaica, Japan, Mauritius, Mexico, Morocco, North Sea bordering nations, Peru, Thailand, most U.S. states)

42. competition between onshore and off-shore fishermen for same stocks — harvesting at rate greater than sustainable yield — decreased fish yields, social conflicts between two groups

(Malaysia)

H. Access to the shoreline and subtidal areas.

43. residential development on the shoreline — blocked or impaired public access to the shore — resentment among local inhabitants, increased recreation pressure on accessible areas, site deterioration, decreased recreational quality

(Australia, Greece, Norway, Sweden, most U.S. states)

44. tourism development of shoreline — blocked or impaired public access to the shore — resentment among local inhabitants, increased recreation pressure on accessible areas, site degradation, decreased recreational quality

(Barbados, Dominican Republic, Greece, Grenada, Jamaica, Trinidad (Tobago), United States (Florida, Hawaii, California), Virgin Islands)

Appendix B

USE OR ACTIVITY	⟶	ENVIRONMENTAL CHANGE	⟶	IMPACT OF SOCIAL CONCERN

I. Visual quality.

45. residential development (particularly second home) | decreased visual quality of rural or natural landscapes | decreased recreation and tourism quality

(Australia, Barbados, Denmark, Finland, France, Greece, New Zealand, Norway, Seychelles, Spain, Sweden, United Kingdom)

46. tourism development | decreased visual quality of rural or natural landscapes | decreased recreation and tourism quality

(Australia, Bahamas, Barbados, Fiji, France, Greece, Jamaica, Israel, Mauritius, Poland, Seychelles, Sri Lanka, Tahiti, Tanzania, Togo, Western Samoa)

J. Employment and cultural values.

47. tourism development | increased salaries in tourism sector relative to other sectors | loss of agricultural workers, decreased agricultural productivity

(Jamaica, Barbados)

48. tourism development | erosion of local customs and cultural values | resentment and social problems among nationals

(Grenada, Jamaica, Mexico, Virgin Islands, Western Samoa)

II. HAZARDS

1. Shoreline erosion (naturally occurring)

(Australia, German Democratic Republic, Guatemala, Japan, Philippines, Spain, Sri Lanka, Togo, United Kingdom, USSR (Black Sea), most U.S. states)

148

2. Coastal river flooding

    (China, El Salvador, Ethiopia, Guatemala, India, Malaysia, Mauritius, Netherlands, Panama, Philippines, Tanzania, Togo, United Kingdom)

3. Storms (wind, wave and water damage)

    (all Caribbean Islands (particularly those where population is concentrated on low lying shoreline), China, Fiji, Indonesia, Mauritius, Mexico, all inhabited Pacific coral atolls, Pakistan, Philippines, United States (Florida, Hawaii, Louisiana, Texas))

4. Tsunamis

    (Ecuador, Indonesia, Pakistan, Venezuela, West Indies, United States (Alaska, California, Hawaii))

5. Migrating dunes (cover infrastructure and/or agriculture)

    (Somalia)

III. SECTORAL PLANNING

1. Fisheries development (particularly conversion of artisanal fisheries)

    (Brazil, Cape Verde Islands, all islands of Commonwealth Caribbean, Guyana, Honduras, Nicaragua, Nigeria, Pakistan, Seychelles, Tanzania)

2. Natural area protection systems (including marine parks)

    (Argentina, Australia, Dominican Republic, Ecuador, Indonesia, Ireland, Kenya, Tanzania, United Kingdom, all U.S. coastal states)

3. Water supply (often a function of overdrafting coastal aquifers)

    (Cape Verde Islands, China (Pearl River Delta), Ethiopia, Guatemala, Guyana, Israel, Morocco (desalinization), Pakistan, Thailand, Togo, Windward and Leeward Islands)

4. Recreation development (primarily for residents)

    (Australia, Ireland, Israel, New Zealand, Norway, Sweden, United Kingdom)

5. Tourism development (particularly potential areas and/or infrastructure needs)

    (Dominican Republic, Mozambique, Seychelles)

6. Energy development (particularly ocean thermal energy conversion (OTEC))

    (Brazil, Chile, Fiji, Hawaii, Sri Lanka, Tonga, Western Samoa)

7. Port development (particularly new ports)

    (Cape Verde Islands, Ethiopia, Guatemala, Japan, Mexico, New Zealand, Singapore, Western Samoa)

8. Oil or toxic spill contingency planning

    (Dominican Republic)

9. Industrial siting (often in conjunction with increasing employment in depressed or impoverished areas)

    (Finland, Greece, Japan, Nigeria, Sweden, most U.S. states)

10. Agricultural development

    (Belize, Indonesia, Kenya, Surinam)

11. Maricultural development (particularly shrimp)

    (Bangladesh, Ecuador, Indonesia, Mexico, Panama, Philippines, Thailand)

IV. COMMENTS

The impact chains presented in Part I of this appendix are a reference list. They are not intended to explain cause and effect relationships. Impact assessment analysts have documented cause and effect relationships in greater detail. However, such rigorous description often produces more information than is needed for an index. Impact issues only need to be described in sufficient detail to distinguish the multiple ways one use activity (whether it is coastal or not) affects a coastal use activity.

We list 48 impact issues. We derived the impact issues from a review of the literature cited in the references section. (A few impact issues known to occur in developed nations were not listed because none of the literature mentioned it as a concern. An example is the adverse visual impact of coastal industrial development, such as port facilities and refineries.)

There are two important limitations in this list of nations. One is the number of national descriptions we located. Descriptions were found for only 76 of the coastal nations -- less than half the world's total. A second problem is the sketchy nature of the descriptions. Adequate information to draw conclusions about issues is available for only about 30 of the 76 coastal nations. Most information on the other 46 coastal nations listed in this appendix comes from **Coastal Area Management and Development** (UNOETB, 1982a), "Coastal Zone Management" (Mitchell, 1982), **Interregional Seminar on Development and Management of Resources of Coastal Areas** (Skekielda and Breuer, 1976), **Man, Land and Sea** (Soysa, Chia, and Collier, 1982), and **Marine and Coastal Area Development in the East African Region** (UNEP, 1982a).

Most lists of coastal issues were an analyst's opinion and were not supported by systematic assessment (such as a review of national literature, interviews, or concerns from a national conference). In such cases one analyst's list might differ from the opinion of another professional. The quality of a list of national issues will depend on the expertise, biases, and knowledge gaps of the individual or the group who makes the compilation.

Coastal resources issues may be very real without gaining a significant place on a nation's political agenda. Many nations -- particularly the lower income nations -- may experience adverse impacts on their fisheries yield or community health without this condition being widely recognized. There may be inadequate information about cause and effect relationships. As we use the term, a problem becomes an issue when the government recognizes its existence and places it on the public policy making agenda.

Determining the relative importance of issues requires that criteria be set. This in turn raises the political question of who should set the priorities. Two criteria commonly used to determine the degree of importance are the number of people affected and the potential monetary benefits to be derived. Clearly, the rating will depend on the perspective and biases of the evaluator. For this reason, an attempt should be made to control bias. One approach would be to delegate the task to a national panel representing the spectrum of coastal resource and environmental users, agencies, and scientists.

Setting priorities for national issues raises the additional question of whether foreign interests should participate in setting the rating of national issues. For example, international conservation and scientific organizations such as IUCN and the Pacific Science Congress, as protectors of the world's natural resource heritage, have a legitimate interest in the environments and fauna of all nations. Both the issues identified and the ratings made by conservation organizations are likely to be quite different from those of some developing nations. For example, preservation of genetic diversity is not likely to be a high priority issue on the public agenda of a developing nation.

151

If professionals in international coastal resources management agree that a global coastal issues index should be prepared, the next step would be to survey coastal nations to identify the issues and rank their relative importance. Several methods could be used to obtain this information, the simplest being a further review of the literature. Our literature search for this report was not exhaustive. Many additional descriptions (particularly in French and Spanish) of coastal nations' environmental programs are likely to include discussions of coastal issues.

Conducting structured surveys is another way to identify and rank issues. We suggest that our global issues list, with some modifications, could be the basis for structuring the survey. This index can serve as a checklist to determine if a coastal nation has a specific problem or need for sectoral development. Criteria and measures should be used to assess the extent of the problems.

Professional institutions such as **Coastal Management Journal**, the Coastal Society, the Marine Technology Society, and the International Geophysical Union could undertake a structured survey among their overseas members or collegial counterparts. International assistance agencies -- particularly USAID -- could survey their national missions. In USAID's case, it seems useful to add a section on coastal management issues to the second phase of their country environmental profile series. This has already been done for the first published phase two report, **The Dominican Republic Country Environmental Profile, A Field Survey**. We hope that this precedent will be continued. If it is, future reports should indicate the relative importance of the coastal issues.

Some of the impact issues should be divided into more specific categories. In this way, the effect of each use activity on each other coastal use could be listed separately. For example, we combined port development and offshore shipping of oil in impact issue no. 7. In turn, the impacts of these issues were combined (decreased fish yields and decreased recreation or tourism quality). In a more thorough treatment, issue no. 7 could become four lists:

o   port development - decreased fish yields;

o   offshore shipping of oil - decreased fish yields;

o   port development - decreased tourism quality; and

o   offshore shipping - decreased tourism quality.

APPENDIX C

AN OUTLINE FOR DESCRIPTIONS OF
COASTAL RESOURCES MANAGEMENT PROGRAMS

Setting and Brief Historical Perspective

Major coastal environments.

Significant coastal resources, value of coastal resources to the nation.

Major, ongoing, coastal resource uses and activities.

Critical problems of coastal use and activity.

Governance Structure

Ownership of coastal areas (particularly relative rights of the public and private sector).

Governance of coastal areas (legal powers, government organization and procedures).

Decision Making

Who makes critical decisions?

What criteria and information are used?

What appear to be the major factors that influence decision making?

How are decisions implemented and enforced?

Evaluation

Critical problems being addressed.

Critical problems not being addressed.

Ability of governance structure and decision making process to address problems.

Major implementation problems experienced or anticipated.

Prospects for change.

APPENDIX D

NEED FOR ISSUE-BASED GOVERNANCE ANALYSIS

Chapter 8 presented basic concepts about the complexity of the governance arrangements in a nation's coastal zone. This classification fills a void in the literature on institutional complexity in the management of coastal resources. Some literature touches on the factors that create complexity. However, most documents simply list the problems caused by specialization and differentiation in government: fragmentation, gaps in sectoral functions, and overlapping and duplicate sectoral functions. Governance arrangements need to be defined and analyzed more thoroughly by each coastal nation considering initiation of an effort to mount an integrated coastal zone management program.

Many states participating in the U.S. coastal management program analyzed their coastal governance arrangement as a first step in program preparation. (This process is often called institutional stock taking.) A number of developed and developing nations such as the United Kingdom, Australia, Canada, the Philippines, Indonesia, Sri Lanka, and Malaysia, have conducted institutional analyses.

Analyses of the governance arrangement should be issue-based. More specifically, the analysis should focus on the most compelling coastal issues that motivate and initiate a coastal management program. Therefore, analyses of governance arrangements should be organized to reflect major issues currently confronting the coastal nation. For each issue, inputs and intervening factors should be identified.

Inputs:

o   the laws and policies that affect the issues;

o   the government units that are mandated to implement these laws and policies and their specific responsibilities.

Intervening Factors: (characteristics of the above inputs which in turn influence the issue)

o   gaps in responsibility (e.g., either no government mandate or a mandate so vague that it cannot be implemented);

o   fragmentation of responsibility among different units of government;

o   overlaps and duplication of effort among competing units of government;

o   conflicts between units of government trying to achieve their respective mandates.

A standardized process would assist coastal nations in conducting issue-based governance analyses. Such an analytic process would also reinforce USAID's program to prepare environmental profiles of nations eligible for its assistance. For example, the Dominican Republic environmental profile (Hartshorn et al., 1981) presents a rich source of environmental information about the Dominican Republic. However, the report does not clearly show the relationship between the issues and the present governance arrangement for the coastal zone.

The environmental profile offers numerous recommendations on various improvements the Dominican Republic could make in laws, governmental arrangements, and management strategies to resolve environmental problems. The content and structure of the analysis, however, is not issue organized and therefore it is difficult -- if not impossible -- to portray an overall set of optional governance arrangements and management strategies that the Dominican Republic could adopt to improve the regulation of its environment.

APPENDIX E

CLASSIFICATION OF COASTAL GOVERNANCE ARRANGEMENTS

There are 136 sovereign and 40 semi-sovereign coastal states. Does this imply that there are there 176 different governance arrangements for the management of coastal resources and environments? Although no two coastal nations will ever have exactly the same governance arrangements, we can identify some major similarities and differences. When governmental tiers and the geographical divisions are added to the array of sectoral and functional divisions, significant differences among nations are bound to occur. Organizing governance arrangements into groupings defined by similar attributes should help advance the international practice of integrated coastal zone management.

Several analysts have suggested that national governance arrangements for the management of coastal resources and environments should be classified consistently (Mitchell, 1982; Englander, Feldman, and Hershman, 1977). There are two persuasive reasons why the international assistance community should formulate such a system of classification:

o   to provide a framework to comparatively assess
     national coastal management efforts;

o   to identify pre-conditions for adoption of program
     components that have met with significant success
     (Mitchell, 1982).

Our literature review identified just one proposed system for classification of similar governance arrangements. Mitchell's (1982) chapter in **Ocean Yearbook 3**, "Coastal Zone Management: A Comparative Analysis of National Programs," suggests three criteria to classify coastal programs:

1.   **Focus:**

      coastal-specific missions to deal with substantive
      systematic problems (Sri Lanka's approach is given
      as the example);

      -- or --

      coastal management as merely one task of an
      agency with broad functional responsibilities such as
      land use planning or national economic development
      (United Kingdom's approach is given as the
      example);

2.   **Strength of national control:**

      relatively weak national control; high levels of
      regional or local governmental control; variable
      program content and opportunities for public

participation (the U.S. approach is given as the example);

-- or --

strong national control; use of formally specified management systems with mandatory components and limited public participation (the French approach is given as the example).

3. **Orientation:**

policy orientation primarily to enhance economic development goals and mitigate national hazards (Japan's approach is given as the example);

-- or --

policy orientation toward environmental preservation and a tendency to stress amenity considerations (the United Kingdom's approach is given as the example).

Our literature review confirms that these dimensions are important factors that shape coastal management programs. They produce an eight division classification as shown in Figure E.1.

The eight classes appear to be a workable set, and examples for most of them can be readily identified. The real test of the classification is whether all the world's coastal nations will readily fit into the category, "nations with coastal specific programs." In fact, the utility of Mitchell's classification is limited. The vast majority of coastal nations do not have a "coastal specific" program, so they do not fit into this framework.

We have identified about seven nations and 25 subnational units that have established programs specifically designed to manage coastal resources and environments in an integrated fashion. Our literature review confirms the fact that a very great majority of the world's coastal nations either:

o   do not have a national, state, or regional program with particular regard for the integrated management of coastal resources or environments,

-- or --

o   regard the integrated management of coastal resources and environments as a component of another governmental program such as land use or environmental planning.

We propose a revised classification to reflect the reality that all coastal nations manage one or more coastal resources. Our classification recognizes

Figure E.1: Mitchell's Typology of Governance Arrangements

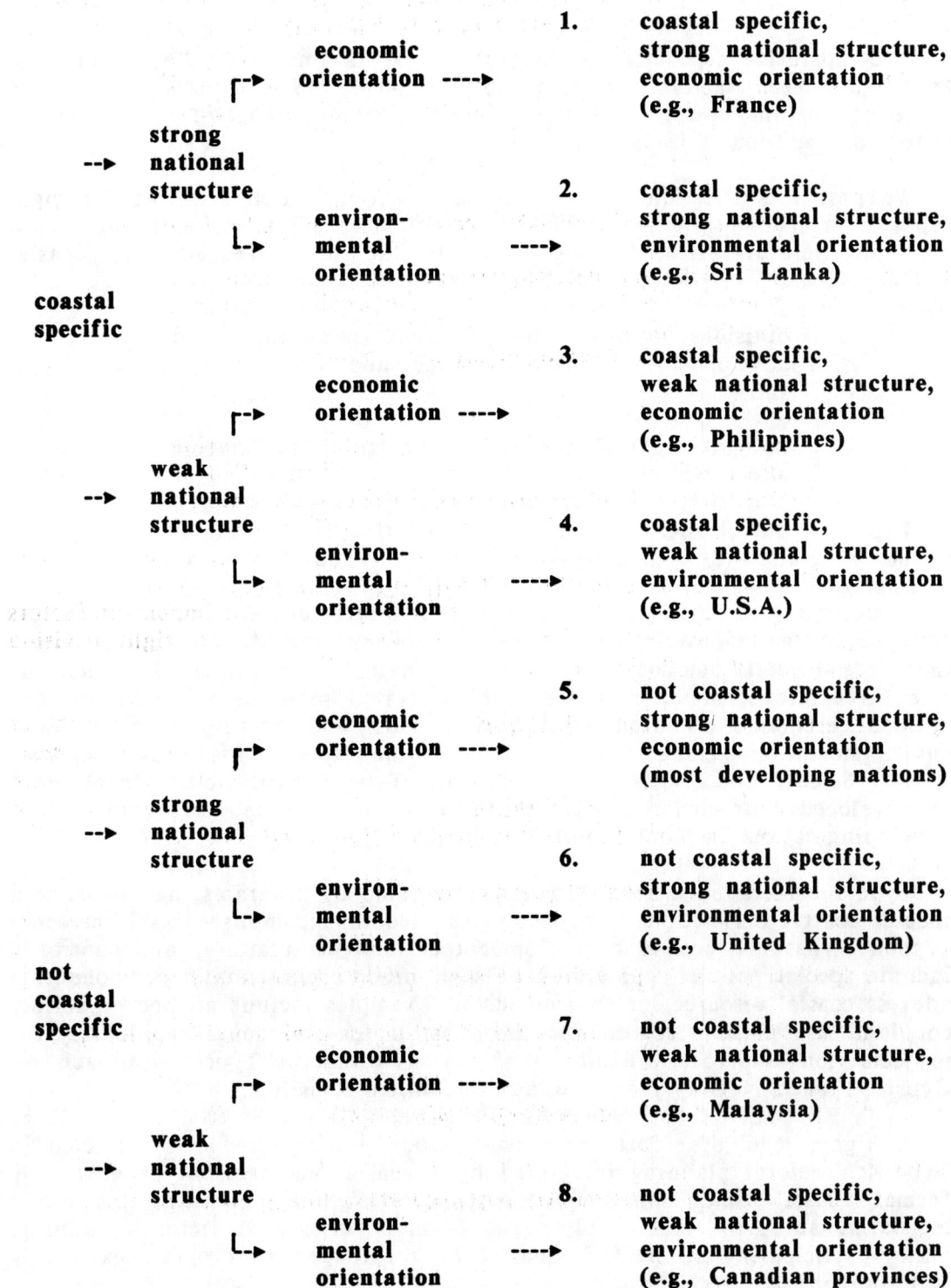

coastal
specific

  --▶ strong
national
structure

      ┌-▶ economic
orientation ----▶

1. coastal specific,
strong national structure,
economic orientation
(e.g., France)

      └-▶ environ-
mental ----▶
orientation

2. coastal specific,
strong national structure,
environmental orientation
(e.g., Sri Lanka)

--▶ weak
national
structure

      ┌-▶ economic
orientation ----▶

3. coastal specific,
weak national structure,
economic orientation
(e.g., Philippines)

      └-▶ environ-
mental ----▶
orientation

4. coastal specific,
weak national structure,
environmental orientation
(e.g., U.S.A.)

not
coastal
specific

--▶ strong
national
structure

      ┌-▶ economic
orientation ----▶

5. not coastal specific,
strong national structure,
economic orientation
(most developing nations)

      └-▶ environ-
mental ----▶
orientation

6. not coastal specific,
strong national structure,
environmental orientation
(e.g., United Kingdom)

--▶ weak
national
structure

      ┌-▶ economic
orientation ----▶

7. not coastal specific,
weak national structure,
economic orientation
(e.g., Malaysia)

      └-▶ environ-
mental ----▶
orientation

8. not coastal specific,
weak national structure,
environmental orientation
(e.g., Canadian provinces)

(Source: Mitchell, 1982)

159

both the global paucity of programs for integrated coastal resources management **and** the ubiquity of non-coastal specific management programs. This classification is illustrated by Table E.1.

A comparative assessment of institutional arrangements should capture and reveal the main factors affecting the ability of the governance process to achieve program objectives (e.g., maintain sustained-yield of a fishery or reduce degradation of resources).

We pose this question: Does the classification which sets out the eight types of arrangements displayed in Figure E.1 meet this criterion? Our review of the literature strongly suggests that the main features of coastal governance are:

o   divisions caused by sectoral planning and
     development of coastal resources and environments;
     and

o   integrated planning efforts to counteract the effects
     of sectoral divisions.

Type 1 in our classification is sectoral planning and development with little or no integration to connect the sectors. Many developing nations fit into this category.

Type 2 is an improvement: sectoral planning integrated by planning efforts that **do not** single out coastal resources or environments for special attention. The three most common strategies for integrated planning of this type are national economic planning, land use or town and country planning, and environmental impact assessment. Japan, the Netherlands, New Zealand, Poland, Sweden and Singapore exemplify the Type 2 institutional arrangement in developed countries. Chile, Fiji, Mexico and Venezuela are examples of developing nations in this category.

Type 3 consists of sectoral planning integrated by programs that **do** make a special coastal distinction. The strategies used to accomplish this integration -- such as national economic development or land use planning and control -- include special policies, guidelines, or some other programmatic component to address coastal resources or environments. Examples include ad hoc guidelines for land use plans prepared for the coast or environmental guidelines for projects along the coast. Examples of nations using the Type 3 approach are Cyprus, France, Norway, Thailand, and the United Kingdom.

A higher level of effort for coastal resources management is reflected in Type 4: sectoral planning integrated by a coastal management program. A formal coastal zone management program, designated by the appropriate legislative authority, is the only major form of integrated sectoral planning. States participating in the U.S. Coastal Zone Management program exemplify this approach.

Finally, Type 5 consists of sectoral planning integrated by a coastal zone management program, and reinforced with another management strategy, such

Table E.1: Coastal Governance Arrangements

Types of Governance Arrangements	1. Sectoral Planning and Development	2. Integrated Planning With <u>No</u> Particular Regard For the Coastal Zone	3. Integrated Planning With Particular Regard For the Coastal Zone	4. Integrated Coastal Zone Management Program
TYPE ONE				
Many, if not most, developing nations	X			
TYPE TWO				
Most developed nations (e.g., Japan, Netherlands, New Zealand, Poland, Sweden, Singapore)	X	X		
Many developing nations (e.g., Chile, Fiji, Mexico, Venezuela)				
TYPE THREE				
(e.g., Cyprus, France, Norway, Thailand, the United Kingdom)	X		X	
TYPE FOUR				
(e.g., United States)	X			X
TYPE FIVE				
(e.g., Brazil, Costa Rica, Ecuador, Greece, Israel, New South Wales, Queensland, Sri Lanka, South Australia)	X	X		X

as national economic development. Brazil, Costa Rica, Ecuador, Greece, Israel, New South Wales, Queensland, South Australia and Sri Lanka are examples.

STEPS IN A FACILITATED POLICY DIALOGUE
OR MEDIATED NEGOTIATION

Facilitated or mediated dispute resolution processes are procedurally complex, yet the many steps in the process are intended to produce agreements that are better informed and more fair, efficient, and durable than "solutions" that might be imposed by a single agency. Here, 11 steps are summarized that might compose mediated negotiation or a policy dialogue (after Fisher and Ury, 1981; Susskind and McCreary, 1985; Susskind and Cruikshank, 1987). Also listed are some questions that may need to be addressed in designing a policy dialogue or mediation process.

1. **Entry of the Nonpartisan Intervenor**

   How is "help" triggered? Who should ask for help? Should the convenor be an expert in mediation techniques, an expert in the issues at hand, or both? Should the nonpartisan party be appointed by a high government official, or even the President or Prime Minister, or should the third party come forward as a volunteer intervenor? Who pays for the services of the third party?

2. **Choosing Representatives**

   Which parties should participate? Only government agencies with conflicting policy goals? Users of coastal resources? Non-governmental organizations? Multinational corporations with interests in coastal resources? Bilateral or multilateral lending institutions? By what criteria are parties determined to have a legitimate stake in the issues? Which spokespersons should represent the interests? In other words, which interests, and which spokespersons should participate in the process?

3. **Setting the Agenda**

   Which policy issues, regulatory standards, or site-specific use conflicts should be discussed? Is there a specific order in which these issues should be taken up? Are certain issues linked in such a way that they should be considered together? What's the schedule for the discussions? Is there a deadline?

   What are the ground rules? Who convenes the meetings and who chairs them? How are uncooperative parties dealt with? What kind of communication is there between negotiators and their constituents during the process?

4. **Joint Fact-finding**

What dimensions of the natural systems and technology are in dispute? What data and analysis might help clarify these issues? Do the parties each have the capability to understand technical material? How can unequal capability be addressed? Is a resource pool needed?

5. **Invention of Options**

What responses can be generated to the problems at hand? Should contingent options be considered to account for different sets of future events? Should the mediator be asked to invent options?

6. **Packaging of Options**

Are there enough issues on the table to make trades possible? Are there interdependent issues under discussion so bargaining can lead to a positive sum outcome? Are parties willing to give something up to get something else in exchange? Should the mediator invent specific compromise proposals? Should parties develop entire competing proposals, or work on single text of an agreement and negotiate each portion of it?

7. **Signing a Written Agreement**

Can negotiators speak for their constituents? Are they willing to sign a negotiated agreement?

8. **Selling the Agreement Among the Constituency**

Did the negotiator for a given interest come away from the table with an agreement acceptable to the people he or she represents? Will it have to be revised slightly to be acceptable "back home"?

9. **Ratification**

Are some last-minute revisions needed to meet the requirements discovered in the previous step? How can a written, but still informal, agreement be linked to more formal mechanisms? What legislation, contracts, covenants, or interagency agreements need to be signed?

10. **Monitoring and Evaluation**

How will parties be held to their written promises? If there were contingent clauses in the agreement, did forecasted events materialize or not? Should the negotiators automatically reconvene after a fixed period of time?

11. **Remediation**

Should an updated agreement be negotiated later if conditions change?

REFERENCES AND BIBLIOGRAPHY

Adams, J. 1973. Proposition 20: A citizen's campaign. **Syracuse Law Review** 24(3):1019-1046.

Adams, T. 1964. **First World Conference on National Parks.** Washington, D.C.: U.S. Department of Interior.

Aderounmu, A. 1976. Nigeria. In **Interregional seminar on development and management of resources of coastal areas,** edited by K. Skekielda and B. Breuer, 213-217. West Berlin: German Foundation for International Development and the United Nations Ocean Economics and Technology Office.

Adulavidhaya, K., Sanit, A., Chindai, S., Rowchai, S., and Rougrujipimon, C. 1982. Thailand: Use and development of mangrove and fisheries resources. In **Man, land, and sea: Coastal resource use and management in Asia and the Pacific,** edited by C.H. Soysa, C.L. Chia, and W.L. Collier, 273-296. Bangkok: Agricultural Development Council.

Ahmed, Y., ed. 1982. **Coastal tourism.** Environmental Management Guidelines Series. Nairobi: United Nations Environment Program.

Akaha, T. 1984. Conservation of the environment of the Seto Inland Sea: A policy response under uncertainty. **Coastal Zone Management Journal** 12(1):83-137.

Al-Abbar, F. 1976. Kuwait. In **Interregional seminar on development and management of resources of coastal areas,** edited by K. Skekielda and B. Breuer, 189-194. West Berlin: German Foundation for International Development and the United Nations Ocean Economics and Technology Office.

Al-Gain, A., Clark, J., and Chiffings, T. 1987. A coastal management program for the Saudi Arabian Red Sea coast. In **Coastal Zone '87, Proceedings of the Fifth Symposium on Coastal and Ocean Management,** vol. 1, 1673-1681. New York: America Society of Civil Engineers.

Allen, T., and Cooney, S. 1973. Institutional roles in technology transfer: a diagnosis of the situation in one small country. **Research and Development Management** 4(1): 41-51.

Allen, T., and Meyer, A. 1982. Technical communication among scientists and engineers in four organizations in Sweden: Results of a pilot study. Working Paper WP 1318-82. Cambridge, MA: Sloan School of Management, Massachusetts Institute of Technology.

Alvarez, J., and Alvarez, S. 1984. **Conceptos basicos sobre manejo costero.** Argentina: Instituo de Pulicaciones Naveales.

Alwis, L. 1982. River basin development and protected areas in Sri Lanka. Paper presented at the World National Parks Congress, October 11-12, 1982, Bali, Indonesia.

References and Bibliography

Amarasinghe, S., and Wickremeratne, H. 1983. The evolution and implementation of legislation for coastal zone management in a developing country: The Sri Lankan experience. In **Coastal Zone '83, Proceedings of the Third Symposium on Coastal and Ocean Management,** vol. 3, 2822-2841. New York: American Society of Civil Engineers.

Amarasinghe, S., Wickremeratne, H., and Lowry, K. 1987. Coastal zone management in Sri Lanka 1978-1986 Appendix. In **Coastal Zone '87, Proceedings of the Fifth Symposium on Coastal and Ocean Management,** vol. 2, 1995-2006. New York: American Society of Civil Engineers.

Amir, S. 1984. Israel's coastal program: Resource protection through management of land use. **Coastal Zone Management Journal** 12(2/3):189-224.

Argentina, government of, the Organization of American States and United Nations Environment Program. 1978. **Environmental quality and river basin development: A model for integrated analysis and planning.** Washington D.C.: Secretary General, Organization of American States.

Armstrong, J., Bissell, H., Davenport, R., Goodman, J., Hershman, M., and Sorensen, J. 1974. **Coastal zone management: The process of program development.** Sandwich, MA: Coastal Zone Management Institute.

ASEAN Experts Group on the Environment. 1983. **General guidelines on the development and management of coastal areas,** Document no. 21. Sixth Meeting of ASEAN Experts Group on the Environment, March 22-24, Jakarta, Indonesia.

Ashbaugh, J., and Sorensen, J. 1976. Identifying the "Public" for participation in coastal zone management. **Coastal Zone Management Journal** 2(4):383-411.

Australia, House of Representatives, Standing Committee on Environment and Conservation. 1980. **Report on management of the Australian coastal zone.** Canberra: Australia Government Publication Service.

Ayon, H. 1988. **Grandes rasgos geomorfologicos de la costa Ecuatoriana.** Ecuador: Fundacion Pedro Vicente Maldonado.

Aziaha, Y. 1976. Togo. In **Interregional seminar on development and management of resources of coastal areas,** edited by K. Skekielda and B. Breuer, 235-242. West Berlin: German Foundation for International Development and the United Nations Ocean Economics and Technology Office.

Bacon, P.R., Deane, C.A., and Putney, A.D. 1988. **A workbook of practical exercises in coastal zone management for tropical islands.** London: Commonwealth Science Council.

Bahamas, government of. 1976. **Planning guidelines for the control of land use and development in the Commonwealth of the Bahamas.** Nassan: Government Printing Office.

Baker, D. 1976. Barbados. In **Interregional seminar on development and management of resources of coastal areas**, edited by K. Skekielda and B. Breuer, 39-61. West Berlin: German Foundation for International Development and the United Nations Ocean Economics and Technology Office.

Baines, G. 1987. Coastal area management in the South Pacific Islands. In **Coastal Zone '87, Proceedings of the Fifth Symposium on Coastal and Ocean Management**, vol. 1, 1682-1695. New York: American Society of Civil Engineers.

Baldwin, M.F. 1988. Environmental laws and institutions of Sri Lanka. An assessment for USAID, Colombo, Sri Lanka.

Banta, J. 1978. The coastal commission and state conflict and cooperation. In **Protecting the golden shore**, edited by R. Healy, 97-131. Washington, D.C.: The Conservation Foundation.

Barnes, J. 1984. Non-governmental organizations: Increasing the global perspective. **Marine Policy** 8(2):171-197.

Beller, W., ed. 1979. **Transactions at the Conference on Environmental Management and Economic Growth in the Smaller Caribbean Islands.** Washington, D.C.: U.S. Department of State.

Biney, C. 1987. Combating coastal pollution in Ghana. In **Coastal Zone '87, Proceedings of the Fifth Symposium on Coastal and Ocean Management**, vol. 1, 960-969. New York: American Society of Civil Engineers.

Bird, E., and Ongkosongo, O. 1981. **Environmental changes on the coasts of Indonesia.** NRTS-12/UNUP-197. Tokyo: The United Nations University.

Birke, P., and Roeseler, W. 1983. State response to the Coastal Zone Management Act of 1972. In **Coastal Zone '83, Proceedings of the Third Symposium on Coastal and Ocean Management**, vol. 1, 206-225. New York: American Society of Civil Engineers.

Biswas, A., and Geping, Q., eds. 1987. **Environmental impact assessment in developing countries.** London: Tycooly.

Blair, D. 1979. **Coastal resource management in the Gulf of Nicoya.** Seattle: Institute for Marine Studies, University of Washington.

Borgese, E., and Ginsburg, N., eds. 1982. **Ocean Yearbook 3.** Chicago: University of Chicago Press.

Boule, M., Olmsted, E., and Miller, T. 1983. **Inventory of wetland resources and evaluation of wetland management in western Washington.** Prepared for the Washington Department of Ecology. Seattle: Shapiro and Associates.

Brazil, Comissao Interministerial para os Recursos do Mar. 1980. **Politica nacional para los recursos do mar.** Brasilia: Interministerial Commission on Ocean Resources.

Brazil, Comissao Interministerial para os Recursos do Mar. 1981. **Ano setorial para los recursos do mar (1982-1985).** Brasilia: Interministerial Commission on Ocean Resources.

Briggs, D., and Hansom, J. 1982. Potential role of ecological mapping in coastal zone management in Europe. **Ekistics** 293 (March-April):114-118.

Broadus, J., Pires, I., Gaines, A., Bailey, C., Knecht, R., and Cicin-Sain. 1984. **Coastal and marine resources management for the Galapagos Islands.** Woods Hole Oceanographic Institution Technical Report WH01-84-43. Woods Hole, MA: Woods Hole Oceanographic Institution.

Broadus, J. 1985. Poor fish of Rendondo! Managing the Galapagos waters. **Oceanus** 28(1):95-99.

Brower, D., and Carol, D., eds. 1987. **Managing land use conflicts: Case studies in special area management.** Durham, NC: Duke University Press.

Brown, L., Brewtun, J.L., Evans, T.J., McGowen, J.H., White, W.A., Groat, C.G., and Fisher, W.L. 1980. **Environmental geologic atlas of the Texas coastal zone: Brownsville-Harlingen area.** Austin, TX: Bureau of Economic Geology, University of Texas at Austin.

Caballero, H. 1976. Monografia de las costas de Honduras. In **Interregional seminar on development and management of resources of coastal areas,** edited by K. Skekielda and B. Breuer, 119-133. West Berlin: German Foundation for International Development and the United Nations Ocean Economics and Technology Office.

California Coastal Zone Conservation Commissions. 1975. **California coastal plan.** Sacramento: State of California Documents and Publications.

California Public Resources Code Section 30000 et. seq. **The Coastal Act of 1976.**

Caldwell, L. 1985. **International environmental policy: Emergency and dimensions.** Durham, N.C.: Duke University Press.

Calvo, G. 1988. The coastal and marine areas of Uruguay. Master's thesis, Nova University, Dania, Florida.

Camhis, M., and Coccossis, H. 1982. The national coastal management program of Greece. **Ekistics** 293:131-138.

Cambers, G. 1987. Coastal zone management programmes in Barbados and Grenada. **In Coastal Zone '87, Proceedings of the Fifth Symposium on Coastal and Ocean Management,** vol. 1, 1384-1394. New York: American Society of Civil Engineers.

Canadian Council of Resources and Environment Ministers. 1979. **Proceedings of the Shore Management Symposium, October 11-15, 1978,** Victoria, British Columbia. Toronto: CCREM.

Carlberg, E., and Grip K. 1982. Coastal policy in Sweden: Use and protection of marine resources. **Ekistics 293:137-142.**

Center for Ocean Management Studies, University of Rhode Island. 1980. **Comparative marine policy: Perspectives from Europe, Scandinavia, Canada, and the United States.** New York: Praeger Special Studies.

Chaguer, A. 1976. Morocco. In **Interregional seminar on development and management of resources of coastal areas,** edited by K. Skekielda and B. Breuer, 201-211. West Berlin: German Foundation for International Development and the United Nations Ocean Economics and Technology Office.

Chapman, V. 1974. Coastal lands management in New Zealand. **Coastal Zone Management Journal** 1(3):333-348.

Chaverri, R. 1989. Coastal management: the Costa Rican Experience. In **Coastal Zone '87, Proceedings of the Fifth Symposium on Coastal and Ocean Management,** vol. 5, 5273-5294. New York: American Society of Civil Engineers.

Chua, T., and Paw, J. 1987. Aquaculture development and coastal zone management in Southeast Asia: Conflicts and complementarity. In **Coastal Zone '87, Proceedings of the Fifth Symposium on Coastal and Ocean Management,** vol. 2, 2007-2021. New York: American Society of Civil Engineers.

Clark, J. 1978. Natural science and coastal planning: The California experience. In **Protecting the golden shore,** edited by R. Healy, 177-208. Washington, D.C.: The Conservation Foundation.

Clark, J. 1983. **Paper on wetlands for the global issues work group, draft.** Washington, D.C.: U.S. National Park Service, International Affairs Unit.

Clark, J., ed. 1985. **Coastal resources management: Development case studies,** Coastal Publication no. 3., Renewable Resources Information Series. Washington, D.C.: U.S. National Park Service and U.S. Agency for International Development.

Clark, J. and McCreary, S. 1987. Estuarine Sanctuaries. In **Managing land-use conflicts: Case studies in special area management,** edited by D. Brower, and D. Carol. Durham, NC: Duke University Press.

Clark, J., McCreary, S., and Snedaker, S. 1988. **Prospects for integrated coastal resources management in West Africa.** Miami, FL: University of Miami.

Coastal Area Management and Planning Network. 1989. The status of integrated coastal zone management: A global assessment. Summary report of a workshop convened at Charleston, South Carolina, July 4-9, Rosensteil School of Marine Sciences, University of Miami.

Coastal States Organization. 1981. Coastal management: A sound investment. A position paper on future funding for the coastal zone management program. Seattle: Coastal States Organization.

Comal, S. 1976. India. In **Interregional seminar on development and management of resources of coastal areas**, edited by K. Skekielda and B. Breuer, 135-141. West Berlin: German Foundation for International Development and the United Nations Ocean Economics and Technology Office.

Conservation Foundation. 1980. **Coastal zone management -- 1980: A context for debate.** Washington, D.C.: Conservation Foundation.

Cragg, S. 1982. **Coastal resources and the UMBOI Logging Project: An environmental impact study.** Port Moresby, Papua, New Guinea: Forest Products Research Center, Office of Forest, Department of Primary Industry.

Craine, L. 1971. Institutions for managing lakes and bays. **Natural Resources Journal** 11(3):523-546.

Cuenya, B. and Hardoy. J. 1981. **Legal, regulatory, and institutional aspects of environmental and natural resource management in developing countries: A country case study of Venezuela.** Washington, D.C.: U.S. Agency for International Development and National Park Service Natural Resource Project.

Cullen, P. 1977. Coastal management in Port Phillip. **Coastal Zone Management Journal** 3(3):291-305.

Cullen, P. 1982. Coastal zone management in Australia. **Coastal Zone Management Journal** 10(3):183-211.

Cullen, P. 1984. The heritage coast programme in England and Wales. **Coastal Zone Management Journal** 12(2/3):225-257.

Cullen, P. 1987. Coastal resource management and planning. **Australian Planner** 25(3):10-12.

Dalfelt, A. 1982. The role of international assistance organizations promoting effective management of protected areas. Paper presented at the World National Parks Congress, October 11-12, 1982, Bali, Indonesia.

Davoren, W. 1982. Tragedy of the San Francisco Bay commons. **Coastal Zone Management Journal** 9(2):111-153.

Dehart, G., and Glazer, M. 1980. Improved coordination for planning and permitting in special areas: A report to the President for the Federal Coastal Program. Second draft.

DuBois, R. 1985. Coastal fisheries management lessons learned from the Caribbean. In **Coastal resources management: Development case studies,** 291-362. Coastal Management Publication No. 3, NPS/AID Series. Washington, D.C.: U.S. National Park Service and U.S. Agency for International Development.

DuBois, R., Berry, L., and Ford, R. 1985. Catchment land use and its implications for coastal resources conservation. In **Coastal resources management: Development case studies,** 444-503. Coastal Management Publication No. 3, NPS/AID Series. Washington, D.C.: U.S. National Park Service and U.S. Agency for International Development

DuBois, R., and Towle, E.L. 1985. Coral harvesting and sand mining management practices. In **Coastal resources management: Development case studies,** 203-283. Coastal Management Publication No. 3, NPS/AID Series. Washington, D.C.: U.S. National Park Service and U.S. Agency for International Development

Duddleson, W. 1978. How the citizens of California saved their coastal management program. In **Protecting the golden shore,** edited by R. Healy, 3-64. Washington, D.C.: The Conservation Foundation.

Dyer, D. 1972. California beach access: The Mexican law and the public trust. **Ecology Law Quarterly** 2:571-611.

Ecuador, Ministerio de Energia y Minas. 1988. **Structure and objectives of a coastal resources management program for Ecuador and a manifesto in support of the program.** Guayaquil: Government of Ecuador, The Coastal Resources Center at the University of Rhode Island, U.S. Agency for International Development.

Ecuador, Armada y Las Naci nes Unidas. 1983. **Ordenacion y desarrollo integral de las zonas costeras.** Proceedings of a seminar convened May 18-27, 1981, Guayaquil, Ecuador.

Elliott, H. 1974. **Second World Conference on National Parks.** Morges, Switzerland: International Union for the Conservation of Nature and Natural Resources.

Englander, E., Feldman, J., and Hershman, M. 1977. Coastal zone problems: A basis for evaluation. **Coastal Zone Management Journal** 3(3):217-236.

Esman, M. 1978. **Developing and managing institutions for industrial development.** New York: United Nations Industrial Development Organization.

Ewen, L. 1983. Institutional problems in the future management of the California coastal resource program. In **Coastal Zone '83, Proceedings of the Third Symposium on Coastal and Ocean Management,** vol. 1, 252-271. New York: American Society of Civil Engineers.

Finn, D. 1983. Conserving the coastal and marine resources of East Africa. In **Coastal Zone '83, Proceedings of the Third Symposium on Coastal and Ocean Management**, vol. 2, 1823-1839. New York: American Society of Civil Engineers.

Fisher, R. and Ury, W. 1981. **Getting to yes.** Boston: Houghton Mifflin.

France, Interministerial Committee for Regional Development and Planning. 1972. **The Picard report.** Paris.

France, Ministry of the Environment. 1980. French coastal policy. **Ekistics** 293:128-130, translated from Notes Vertes No. 13, **Actions Nouvelles** (November 1980).

Gamman, J., and McCreary, S. 1988. Suggestions for integrating EIA and economic development in the Caribbean region. **Environmental Impact Assessment Review** 8(1):43-60.

Gamman, J., Towers, S., and Sorensen, J. 1974. **Federal involvement in the California coastal zone: A topical index to agency responsibility.** San Diego: Institute of Marine Sciences, University of California.

Gamman, J., Towers, S. and Sorensen, J. 1975. **State involvement in the California coastal zone: A topical index to agency responsibility.** San Diego: Institute of Marine Resources, University of California.

Garnier, E. 1976. Haiti. In **Interregional seminar on development and management of resources of coastal areas,** edited by K. Skekielda and B. Breuer, 113-117. West Berlin: German Foundation for International Development and the United Nations Ocean Economics and Technology Office.

Gentry, C. 1982. **Small scale beekeeping.** Peace Corps Information Collection and Exchange, Manual M-17. Washington, D.C.: U.S. Government Printing Office.

Goldberg, E. 1976. **The health of the ocean.** Paris: UNESCO.

Goodland, R. 1981. **Indonesia's environmental progress in economic development.** Bogor, Indonesia: Center for Natural Resource Management and Environmental Studies, Bogor Agricultural University.

Goodland, R. 1982. Environmental requirements of the World Bank, including wetland conservation. Paper presented at the World National Parks Congress, October 11-12, 1982, Bali, Indonesia.

Great Britain, Countryside Commission. 1970. **The planning of the coastline.** A report of a study of coastal preservation and development in England and Wales. London: Her Majesty's Printing Office.

Grenell, P. 1988. The once and future experience of the California Coastal Conservancy. **Coastal Management Journal** 16(1):13-21.

Gruppe, H., and Ofosu-Amaah, W. 1981. **Legal, regulatory and institutional aspects of environmental and natural resource management in developing countries: A country study of Ghana.** Washington, D.C.: U.S. Agency for International Development and National Park Service Natural Resource Project.

Gruppe, H., and Ofosu-Amaah, W. 1981. **Legal, regulatory, and institutional aspects of environmental and natural resources management in developing countries: A country study of Malaysia.** Washington, D.C.: U.S. Agency for International Development and National Park Service Natural Resource Project.

Gusman, S. 1981. Policy Dialogue. **Resolve Newsletter.**

Gwynne, M. 1982. The global environmental monitoring system (GEMS) of UNEP. **Environmental Conservation** 9(1):35-41.

Hall, D. 1987. Maritime planning in New Zealand. In **Coastal Zone '87, Proceedings of the Fifth Symposium on Coastal and Ocean Management,** vol. 1, 30-44. New York: American Society of Civil Engineers.

Hamilton, L. and Snedaker, S. 1984. **Mangrove management handbook.** Honolulu: The East-West Center.

Hanson, A., and Koesoebiono. 1979. Settling coastal swamplands in Sumatra: A case study for integrated resource management. In **Developing economics and the environment: The Southeast Asian experience,** 121-175. New York: McGraw-Hill.

Harrison, P., and Sewell, W. 1979. Shoreline management: The French approach. **Coastal Zone Management Journal** 5(3):61-180.

Harrison, P., and Parkes, M., eds. 1983. Theme issue: Coastal management in Canada. **Coastal Zone Management Journal** 11(1/2):148.

Hartshorn, G., Antonini, G., Dubois, R., Harcharik, Heckadon, S., Newton, H., Quesada, C., Shores, J., and Staples, G. 1981. **The Dominican Republic: Country environmental profile, a field study.** Prepared for U.S. Agency for International Development. Washington, D.C.: U.S. Government Printing Office.

Hayden, B., Dolan, R., and Ray, G. 1982. A system of marine biophysical provinces for conservational purposes. Paper presented at the World National Parks Congress, October 11-12, 1982, Bali, Indonesia.

Hayden, W., and Dolan, R. 1976. **Classification of coastal environments of the world.** Charlottesville, VA: Department of Environmental Sciences, University of Virginia.

Hayes, M. 1985. Beach Erosion. In **Coastal resources management: Development case studies,** 67-190. Coastal Management Publication No. 3, NPS/AID Series. Washington, D.C.: U.S. National Park Service and U.S. Agency for International Development

Healy, R., ed. 1978. **Protecting the golden shore: Lessons from the California Coastal Commissions.** Washington, D.C.: The Conservation Foundation.

Hershman, M. 1980. Coastal zone management in the United States: A characterization. In **Comparative marine policy: Perspectives from Europe, Scandinavia, Canada, and the United States,** 57-64. New York: Praeger Special Studies.

Herz, R. 1987. A regional program on coastal monitoring and management of mangroves in Brazil. In **Coastal Zone '87, Proceedings of the Fifth Symposium on Coastal and Ocean Management,** vol. 2, 2262-2268. New York: American Society of Civil Engineers.

Hewson, C., Malele, F., and Muller, P. 1979. Coastal area management in Western Samoa. In **Proceedings of the Workshop on Coastal Area Development and Management in the Pacific,** edited by M. Valencia, 113-115. Honolulu: East-West Center and the University of Hawaii.

Hildreth, R. 1975. Coastal land use control in Sweden. **Coastal Zone Management Journal** 2(1):1-29.

Hildreth, R. 1987. EEZ governance in Australia, Canada, The United States, and New Zealand. In **Coastal Zone '87, Proceedings of the Fifth Symposium on Coastal and Ocean Management,** vol. 4, 3414-3429. New York: American Society of Civil Engineers.

Hildreth, R., and Johnson, R. 1983. Coastal zone management on the west coast: An evaluation. In **Coastal Zone '83, Proceedings of the Third Symposium on Coastal Ocean Management,** vol. 3, 231-251. New York: American Society of Civil Engineers.

Hillary, A. 1987. The International Network: A solution for marine protected area management. In **Coastal Zone '87, Proceedings of the Fifth Symposium on Coastal and Ocean Management,** vol. 3, 3601-3605. New York: American Society of Civil Engineers.

Holland, M. 1982. Kuwait waterfront project: An interdisciplinary approach to design. In **Proceedings of the Seventh Annual Conference of the Coastal Society,** 53-60. October 11-14, 1981, Galveston.

Horberry, J. 1983. **Environmental guidelines survey.** Washington, D.C.: International Union for Conservation of Nature and Natural Resources and the International Institute for Environment and Development.

Horberry, J. 1984. Status and application of environmental impact assessment for development. Draft paper. Gland, Switzerland: Conservation Development Center, International Union for the Conservation of Nature and Natural Resources.

Horberry, J., and Johnston, B. 1981. **Environmental performance of consulting organizations in development aid.** Washington, D.C.: International Institute for Environment and Development.

Hulm, P. 1983. **A strategy for the seas: The Regional Seas Programme past and future.** Nairobi: United Nations Environment Program.

Hultmark, E. 1982. On effective coordination of marine resources development within new areas. Working paper prepared for the Expert Group Meeting on Institutional Arrangements for Marine Resource Development. New York: United Nations.

Ibe, A. 1987. Collective response to erosion along the Nigerian coastline. In **Coastal Zone '87, Proceedings of the Fifth Symposium on Coastal and Ocean Management,** vol. 1, 741-754. New York: American Society of Civil Engineers.

Indonesia National Planning Agency and the Canadian International Development Agency. 1988. **Action plan for sustainable development of Indonesia's marine and coastal resources.** Jakarta: Indonesia National Planning Agency.

Inoue, S. 1984. Urban coastal zone and port development in Japan. **Coastal Zone Management Journal** 12(1):57-83.

International Institute for Environment and Development. 1981. **Legal, regulatory, and institutional aspects of environmental and natural resource management in developing countries.** Washington, D.C.: U.S. Agency for International Development and U.S. National Park Service Project.

International Union for the Conservation of Nature and Natural Resources. 1980. **World conservation strategy: Living resources for sustainable development.** Commission on Ecology Papers, Number 3. Prepared in cooperation with United Nations Environment Program. Gland, Switzerland.

International Union for the Conservation of Nature and Natural Resources. 1982. **IUCN's contribution to the formulation of the Action Plan for the Protection and Development of the Marine and Coastal Environment of the East Africa Region.** Submitted to UNEP's Regional Seas Activity Center, Gland, Switzerland.

Ireland, Bord Failte Eireann and An Foras Forbatha. 1972. **National coastline study,** 3 volumes. Dublin: An Foras Forbatha.

Israel, Ministry of the Interior Planning Section. 1978. **The national outline scheme for the Mediterranean coast.** Tel Aviv.

IUCN. **See** International Union for the Conservation of Nature and Natural Resources.

Juda, L. 1987. The exclusive economic zone and ocean management. **Ocean Development and International Law** 18:305-331.

Juda, L., and Burroughs, R. 1990. The prospects for comprehensive ocean management. **Marine Policy** 14:23-35.

Jakobsen, P., Tougaard, N., and Larsen, K. 1987. Copenhagen metropolitan region-coast erosion management. In **Coastal Zone '87, Proceedings of the Fifth Symposium on Coastal and Ocean Management,** vol. 2, 1659-1672. New York: American Society of Civil Engineers.

Jayewardene, H. 1983. The National Aquatic Resources Agency: The Sri Lankan experience. Working paper prepared for the Expert Group Meeting on Institutional Arrangements for Marine Resource Development, January 10-14, 1983. New York: United Nations.

Johnson, B., and Ofosu-Amaah, W. 1981. **Legal, regulatory, and institutional aspects of environmental and natural resource management in developing countries: A country study of Sudan.** Washington, D.C.: U.S. Agency for International Development and National Park Service Project.

Ketchum. B., ed. 1972. **The water's edge: Critical problems of the coastal zone.** Cambridge, MA: Massachusetts Institute of Technology Press.

Kildow, J. 1977. **International transfer of marine technology: A three volume study.** Massachusetts Institute of Technology Sea Grant Program Report, MITSG 77-20. Cambridge, MA: Massachusetts Institute of Technology Press.

Kim, S. 1979. Current activities in coastal area development and management in the Republic of Korea. In **Proceedings of the Workshop on Coastal Area Development and Management in the Pacific,** edited by M. Valencia, 103-110. Honolulu: East-West Center and the University of Hawaii.

Kinsey, D., and Sondheimer, C. 1984. **Country assessments of Ecuador, Thailand, Sri Lanka, Indonesia and Philippines.** Prepared for the U.S. Agency for International Development and U.S. National Ocean Service Program on Coastal Resources Management in Developing Nations.

Kintanar, R. 1987. Coastal zone management of Puerto Galera. In **Coastal Zone '87, Proceedings of the Fifth Symposium on Coastal and Ocean Management,** vol. 1, 2925-2938. New York: American Society of Civil Engineers.

Klapp, M. 1984. A comparative analysis of autonomous state enterprises in oil-industrializing developed and less developed countries. APSA Annual Meeting.

Knecht, R. 1983. Personal communication on Colombia's approach to marine and coastal resources management, September 16, 1983.

Knecht, R., Cicin-Sain, B., Broadus, J., Silva, M., Bowen, R., Marcus H., and Petersen, S. 1984. **The management of ocean and coastal resources in Colombia: An assessment.** Woods Hole Oceanographic Institution Technical Report. Woods Hole, MA: Woods Hole Oceanographic Institution.

Koekebakker, P., and Peet. G. 1987. Coastal Zone Planning and Management in the Netherlands. **Coastal Management Journal** 15(2):121-134.

Koesoebiono, Collier, W., and Burbridge, P. 1982. Indonesia: Resource's use and management in the coastal zone. In **Man, land, and sea**, edited by C. Soysa, L.S. Chia, and W.L. Collier, 115-133. Bangkok: Agricultural Development Council.

Kux, M. 1983. Memorandum on a proposed project on coastal resources management in Indonesia, March 13-14, Jakarta, Indonesia.

League of Women Voters of Washington. 1983. **Public perception of the Washington Shoreline Management Act.** Prepared for the Washington State Department of Ecology. Olympia, WA: League of Women Voters.

Lemay, M. 1987. Common priorities for the management of marine protected areas: Perspectives and results from an international seminar. In **Coastal Zone '87, Proceedings of the Fifth Symposium on Coastal and Ocean Management**, vol. 1, 3587-3600. New York: American Society of Civil Engineers.

Lewis, S. 1975. Coastal plan runs aground. **Planning** 41(November):12-17.

Looi, C. 1987. Coastal zone management plan development in Malaysia: Issues and problems. In **Coastal Zone '87, Proceedings of the Fifth Symposium on Coastal and Ocean Management**, vol. 1, 4601-4615. New York: American Society of Civil Engineers.

Lowry, K. and Wickremeratne, H. 1989. Coastal management in Sri Lanka. In **Ocean Yearbook 7**, edited by E. Borgese, N. Ginsburg, and J. Morgan, 263-293. Chicago: University of Chicago Press.

Maembe, T. 1976. United Republic of Tanzania. In **Interregional seminar on development and management of resources of coastal areas**, edited by K. Skekielda and B. Breuer, 243-255. West Berlin: German Foundation for International Development and the United Nations Ocean Economics and Technology Office.

Mahmud, S. 1985. Impacts of river flow changes on coastal ecosystems. In **Coastal resources management: Development case studies**, 511-583. Coastal Management Publication No. 3, NPS/AID Series. Washington, D.C.: U.S. National Park Service and U.S. Agency for International Development

Marine Law Institute. 1988. **Guidebook to the economics of waterfront planning and water dependent uses.** Portland, ME: University of Southern Maine.

Mazmanian, D., and Sabatier, P. 1983. **Implementation and public policy.** San Francisco: Scott, Foresman and Company.

McAlister, I., and Nathan, R. 1987. Malaysian National Coastal Erosion Study. In **Coastal Zone '87, Proceedings of the Fifth Symposium on Coastal and Ocean Management**, vol. 1, 45-55. New York: American Society of Civil Engineers.

McCrea, M. 1980. Evaluation of Washington State's coastal management program through changes in port development. PhD dissertation, University of Washington.

McCrea, M., and Feldman, J. 1977. Interim assessment of Washington State shoreline management. **Coastal Zone Management Journal** 3(2):119-150.

McCreary, S. 1982. Legal and institutional opportunities and constraints in wetland restoration. In **Wetland restoration and enhancement in California**, edited by M. Josselyn, 39-47. La Jolla, CA: Tiburon Center for Environmental Studies and California Sea Grant Program.

McCreary S. 1985. **The recruitment and application of scientific information in coastal and marine resources management: Analogs to the Galapagos Islands.** Woods Hole Oceanographic Institution Technical Report, WHOI 85-14. Woods Hole, MA: Woods Hole Oceanographic Institution.

McCreary, S. 1987. Facilitated dialogue produces consensus on tributyl tin risks to the marine environment. **Environmental Impact Assessment Review** 7(1):89-92.

McCreary, S. 1989. Resolving Science Intensive Public Policy Disputes: Lessons from the New York Bight Initiative. Ph.D. dissertation prepared for the Department of Urban Studies and Planning, Massachusetts Institute of Technology.

McCreary, S. and Adams, M. 1988. Prospects for transfer of the California Coastal Conservancy model for habitat restoration to other coastal states. **Coastal Management Journal** 16(1):69-92.

McCreary, S. and Robin, R. 1985. The California Coastal Conservancy experience in wetland protection. **Proceedings of the First Annual Conference of the Association of State Wetland Managers**, September 21-24, 1984, Gainesville, Florida.

McCreary, S. and Zentner, J. 1983. Innovative estuarine restoration and management. In **Coastal Zone '83, Proceedings of the Third Symposium on Ocean and Coastal Resources**, vol. 3, 2527-2551. New York: American Society of Civil Engineers.

McGoodwin, J. 1986. The tourism-impact syndrome in developing coastal communities: A Mexican Case. **Coastal Zone Management Journal** 14(1/2):131-146.

McNeeley, J., and Miller, K. 1983. IUCN, national parks and protected areas: Priorities for action. **Environmental Conservation**, 10(1):13-21.

Meltzoff, S., and LiPuma, E. 1986. The social and political economy of coastal zone management: Shrimp mariculture in Ecuador. **Coastal Zone Management Journal** 14(4):349-380.

Mitchell, C. and Gold, E. 1982. **The integration of marine space in national development strategies of small island states: The case of the Caribbean states of Grenada and St. Lucia.** Halifax, Nova Scotia: Dalhousie Ocean Studies Program.

Mitchell, J. 1982. Coastal zone management: A comparative analysis of national programs. In **Ocean Yearbook 3**, edited by E. M. Borgese and N. Ginsburg, 258-319. Chicago: The University of Chicago Press.

Mitchell, J. 1984. Written comments on draft of Sorensen, J., McCreary, S. and Hershman, M., **Institutional arrangements for integrated coastal resources management.** February 1984.

Moran, J. 1976. Ordinacion of aprovechamiento de los recursos de la zona costera de la Republica del El Salvador. In **Interregional seminar on development and management of resources of coastal areas**, edited by K. Skekielda and B. Breuer, 39-61. West Berlin: German Foundation for International Development and the United Nations Ocean Economics and Technology Office.

Morris, M., ed. 1988. **North-south perspectives on marine policy.** Boulder, CO: Westview Press, Inc.

Mumba, J. 1976. Kenya. In **Interregional seminar on development and management of resources of coastal areas**, edited by K. Skekielda and B. Breuer, 177-181. West Berlin: German Foundation for International Development and the United Nations Ocean Economics and Technology Office.

Myers, N. 1980. **The sinking ark: A new look at the problem of disappearing species.** Oxford: Pergamon Press.

Nagao, Y., Kuroda, K., and Kanai, M. 1987. Port re-creation for coastal zone utilization. In **Coastal Zone '87, Proceedings of the Fifth Symposium on Coastal and Ocean Management**, vol. 1, 56-68. New York: American Society of Civil Engineers.

Nanda, V., and Ris, W. 1972. The public trust doctrine: A viable approach to international environmental protection. **Ecology Law Quarterly** 5:291-303.

Neuwirth, D., and Furney-Howe, S. 1983. Coast effectiveness: A better beach for your buck. In **Coastal Zone '83, Proceedings of the Third Symposium on Coastal and Ocean Management**, vol. 1, 930-948. New York: American Society of Engineers.

New York Times. 1985. April 28.

Ngee, L. 1979. Coastal development in Singapore. In **Proceedings of the Workshop on Coastal Area Development and Management in the Pacific,** edited by M. Valencia, 89-93. Honolulu: East-West Center and the University of Hawaii.

Nir, Y. 1976. The Israel Mediterranean coasts. In **Interregional seminar on development and management of resources of coastal areas,** edited by K. Skekielda and B. Breuer, 159-166. West Berlin: German Foundation for International Development and the United Nations Ocean Economics and Technology Office.

Norton, R. 1982. The mediation of ethnic conflict: Comparative implications of the Fiji case. **Journal of Commonwealth and Comparative Politics** 11(3):309-328.

Obayomi, L. 1976. Benin. In **Interregional seminar on development and management of resources of coastal areas,** edited by K. Skekielda and B. Breuer, 63-80. West Berlin: German Foundation for International Development and the United Nations Ocean Economics and Technology Office.

Ochoa, E., Mac as, W., and Marcos, J. 1987. **Ecuador: Perfil de sus recursos costeros.** Guayaquil, Ecuador: Fundacion Pedro Vicente Maldonado.

Odell, R. 1972. **The saving of San Francisco Bay.** Washington, D.C.: The Conservation Foundation.

Olembo, R. 1982. UNEP and protected areas concerns. Paper presented at the World National Parks Congress, October 11-12, 1982, Bali, Indonesia.

Olsen, S., 1987. Report: A collaborative effort in developing the Integrated Coastal Resources Management Project for Ecuador. **Coastal Management** 15(1):97-101.

Ongkosongo, O. 1979. Human activities and their environmental impacts on the coasts of Indonesia. In **Proceedings of the Workshop on Coastal Area Development and Management in the Pacific,** edited by M. Valencia, 67-74. Honolulu: East-West Center and the University of Hawaii.

Oregon, Land Conservation and Development Commission. 1976. Oregon coastal management program. Draft. Salem, OR: Land Conservation and Development Commission.

Palaganas, V., Tambasen, V., and Antonio, V. 1987. The role of the Marine Park Management in the Conservation of Philippine Coastal Resources. In **Coastal Zone '87, Proceedings of the Fifth Symposium on Coastal and Ocean Management,** vol. 3, 3580-3586. New York: American Society of Civil Engineers.

Perez, E. 1988. **Elementos legales y administrativos del manejo de recursos costeros en la Republica del Ecuador.** Quito, Ecuador: Fundacion Pedro Vicente Maldonado.

Perez, E., Robadue, D., and Olsen, S. 1987. Controls on the development and operation of shrimp farms in Ecuador. In **Coastal Zone '87, Proceedings of the Fifth Symposium on Coastal and Ocean Management,** vol. 4, 4476-4487. New York: American Society of Civil Engineers.

Petrillo, J. 1988. The conservancy concept. **Coastal Management Journal** 16(1):1-12.

Philippines, National Environmental Protection Council. 1978. Proceedings of a planning workshop for coastal zone management, Manilla.

Pongsuwan, U. 1976. Thailand. In **Interregional seminar on development and management of resources of coastal areas,** edited by K. Skekielda and B. Breuer, 229-234. West Berlin: German Foundation for International Development and the United Nations Ocean Economics and Technology Office.

Pontecorvo, G., Wilkenson, M., Anderson, R., and Holdowsky, M. 1980. Contribution of the ocean sector to the United States economy. **Science** 208:1000-1006.

Popper, F., 1981. **The politics of land-use reform.** Madison: The University of Wisconsin Press.

Pravdic, V. 1981. **GESAMP, the first dozen years.** Nairobi: United Nations Environment Program.

Raiffa, H. 1983. Mediation of conflicts. **American Behavioral Scientist** 27(2):195-210.

Ramirez, J. 1976. Monographia de Mexico. In **Interregional seminar on development and management of resources of coastal areas,** edited by K. Skekielda and B. Breuer, 195-200. West Berlin: German Foundation for International Development and the United Nations Ocean Economics and Technology Office.

Real Estate Research Corporation. 1977. **Coastal management consequences: Preliminary conclusions and evaluation approach.** Prepared for the U.S. Office of Coastal Zone Management and the U.S Council on Environmental Quality. Washington, D.C.: U.S. Government Printing Office.

Reeson, P. 1976. Monograph on Jamaica. In **Interregional seminar on development and management of resources of coastal areas,** edited by K. Skekielda and B. Breuer, 167-176. West Berlin: German Foundation for International Development and the United Nations Ocean Economics and Technology Office.

Richmond, R. 1979. Fiji. In **Proceedings of the Workshop on Coastal Area Development and Management in the Pacific,** edited by M. Valencia, 103-110. Honolulu: East-West Center and the University of Hawaii.

Riddell, R. 1981. **Ecodevelopment.** New York: St. Martin's Press.

Ridgway, J. 1979. Coastal area development and management in the Solomon Islands. In **Proceedings of the Workshop on Coastal Area Development and Management in the Pacific**, edited by M. Valencia, 103-105. Honolulu: East-West Center and the University of Hawaii.

Rodgers, K. 1984. **Integrated regional development planning: Guidelines and case studies from OAS experience.** Department of Regional Economic Development, Organization of American States in Cooperation with U.S. National Park Service and U.S. Agency for International Development. Washington, D.C.: Organization of American States.

Rosenbaum, N. 1979. Enforcing wetlands regulations. In **Wetland functions and values: The state of our understanding.** Minneapolis: American Water Resources Association.

Ruangchotivit, T. 1979. A coastal land development project. In **Proceedings of the Workshop on Coastal Area Development and Management in the Pacific**, edited by M. Valencia, 101-102. Honolulu: East-West Center and the University of Hawaii.

Ruddle, K. 1982. Environmental pollution and fisheries resources in Southeast Asian coastal waters. In **Man, land, and sea: Coastal resource use and management in Asia and the Pacific**, edited by C.H. Soysa, C.L Chia, and W.L. Collier, 16-37. Bangkok: Agricultural Development Council.

Sabatier, P. 1977. State review of local land-use decisions: The California Coastal Commissions. **Coastal Zone Management Journal** 3(3):255-290.

Sadacharan, D. and Lowry Jr., K. 1987. Resolving fisheries conflicts in Sri Lanka's coastal zone. In **Coastal Zone '87, Proceedings of the Fifth Symposium on Coastal and Ocean Management**, vol. 1, 1982. New York: American Society of Civil Engineers.

Saeijs, H., and de Jong, A. 1982. The ostirschelde and protection of the environment. **Ekistics** 293:150-156.

Saenger, P., Hegerl, E., and Davie, J. 1983. **Global status of mangrove ecosystems.** Gland, Switzerland: International Union for the Conservation of Nature and Natural Resources.

Salm, R., and Clark, J. 1985. **Marine and coastal protected areas: A guide for planners and managers.** Gland, Switzerland: International Union for the Conservation of Nature and Natural Resources.

Salm, R., and Dobbin, J. 1987. A coastal zone management strategy for the Sultanate of Oman. In **Coastal Zone '87, Proceedings of the Fifth Symposium on Coastal and Ocean Management**, vol. 1, 97-106. New York: American Society of Civil Engineers.

Sanchez, E. 1976. Spain. In **Interregional seminar on development and management of resources of coastal areas,** edited by K. Skekielda and B. Breuer, 39-61. West Berlin: German Foundation for International Development and the United Nations Ocean Economics and Technology Office.

Schneider, P. and Tohn, E. 1985. Success in negotiating environmental regulations. **Environmental Impact Assessment Review** 5(1):67-77.

Schwartz, M., and Cambers, G. 1987. The Unesco-Lesser Antilles Coastal Zone Management and Beach Stability Program. In **Coastal Zone '87, Proceedings of the Fifth Symposium on Coastal and Ocean Management,** vol. 1, 24-29. New York: American Society of Civil Engineers.

Scott, J. 1983. **An evaluation of public access to Washington's shorelines since passage of the Shoreline Management Act of 1971.** Olympia: Washington State Department of Ecology.

Shapiro, H. 1984a. Written comments on draft of J. Sorensen, S. McCreary and M. Hershman, **Institutional arrangements for integrated coastal resources in management,** March 9.

Shapiro, H. 1984b. Coastal area management in Japan: An overview. **Coastal Zone Management Journal** 12(1):19-56.

Shapiro, H. 1987. Still more of Japan's Inland Sea Coastal Citizens' Movement. In **Coastal Zone '87, Proceedings of Fifth Symposium on Coastal and Zone Management,** vol. 2, 2197-2208. New York: American Society of Civil Engineers.

Silva, M., and Desilvestre, I. 1986. Marine and coastal protected areas in Latin America: A preliminary assessment. **Coastal Zone Management Journal** 14(4):311-348.

Siddall, S., Atchue III, J., and Murray Jr., R. 1985. Mariculture development in mangroves: A case study of the Philippines, Ecuador and Panama. In **Coastal resources management: Development case studies,** 1-53. Coastal Management Publication No. 3, NPS/AID Series. Washington, D.C.: U.S. National Park Service and U.S. Agency for International Development

Simons, R. 1982. Ten years later: The Smithsonian international experience since the Second World Parks Congress. Paper presented at World National Parks Congress, October 11-12, 1982, Bali, Indonesia.

Skekielda, K., and Breuer, B., eds. 1976. **Interregional seminar on development and management of resources of coastal areas.** West Berlin, Hamburg, Kiel, and Cuxhaven, May 31, to June, 14, 1976. West Berlin: German Foundation for International Development and the United Nations Ocean Economics and Technology Office.

Snedaker, S. 1982. A perspective on Asia mangroves. In **Man, land, and sea: Coastal resource use and management in Asia and the Pacific**, edited by C. Soysa, L.S. Chia, and W.L. Collier, 65-74. Bangkok: Agricultural Development Council.

Snedaker, S. 1983. Personal Communication to USAID, Science and Technology, Forestry and Natural Resources Branch.

Snedaker, S., Dickenson, J.C., Brown, M.S., and Lahmann, E.J. 1986. **Shrimp pond siting and management alternatives in mangrove ecosystems in Ecuador.** Miami, FL: U.S. Agency for International Development.

Snedaker, S. and Getter, C. 1985. **Coastal resource management guidelines.** Coastal Publication No. 2., Renewable Resource Information Series. Washington, D.C.: U.S. National Park Service and U.S. Agency for International Development.

Soegiarto, A. 1976. Indonesia. In **Interregional seminar on development and management of resources of coastal areas**, edited by K. Skekielda and B. Breuer, 39-61. West Berlin: German Foundation for International Development and the United Nations Ocean Economics and Technology Office.

Soegiarto, A. 1983. Marine sciences in Indonesia: Problems and prospects for national development. Paper presented at the Indonesia-U.S. Seminar on Science and Technology, October 3-5, 1983, Washington, D. C.

Sorensen, J. 1978. **State-local collaborative planning: A growing trend in coastal zone management.** Prepared for the U.S. Department of Commerce, Office of Coastal Zone Management.

Sorensen, J. 1990. An assessment of Costa Rica's coastal management program. **Coastal Management Journal** 18(1):37-63.

Sorensen, J. Forthcoming. Comparative analysis of the zona publica in Latin America. First draft.

Sorensen, J., and Brandani, A. 1987. An overview of coastal management efforts in Latin America. **Coastal Management Journal** 15(1):1-25.

Sorensen, J., McCreary, S., and Hershman, M. 1984. **Institutional arrangements for management of coastal resources.** Coastal Publication No. 1, Renewable Resource Information Series. Washington, D.C.: U.S. National Park Service and U.S. Agency for International Development.

Sorensen, J., and McCreary, S. Forthcoming. Resolution of coastal and marine use conflicts in the developing world. In **Coastal and marine multiple use conflicts, problems and approaches** (title tentative). New York: U.N. Office of Ocean Affairs and Law of the Sea.

Soysa, C., Chia, C., and Collier, W., eds. 1982. **Man, land, and sea: Coastal resource use and management in Asia and the Pacific.** Bangkok: Agricultural Development Council.

How To Turn Your Work Group Into A Winning Team

A Powerful One-day Seminar Providing the Management Techniques and Skills That Will Turn Your Work Group Into a Successful, Super-Productive, High-morale Team!

- Be more than a manager – Learn how to be a "coach" and capture the admiration and respect of everyone on your team.

- Experience a powerful team synergy in the biggest and smallest group undertakings.

- Learn the team dynamics that establish an atmosphere of mutual backup and support – with every team member helping each other.

- Encourage team goals above individual interests – powerful solutions that work with even the most difficult, uncooperative employees.

- Expert tips that enable you to conduct team meetings that boom with productivity and enthusiasm.

- "It's not my job." Dozens of simple strategies that assure you'll never again struggle with this common attitude.

- How to foster a positive and spirited team attitude that encourages everyone to give 110%.

- Leave the seminar with a specific action plan that will immediately boost your team's effectiveness.

ONLY $99
CALL TOLL-FREE
1-800-255-6139

FRED PRYOR SEMINARS

A Division of Pryor Resources, Inc.

Copyright ©1990 Pryor Resources, Inc.

The Team Leader's Success Kit: Twenty Essential Tools

1. Learn how a team can make group decisions – while avoiding the frustrations of "decision by committee."
2. Sensible routines that ensure all members understand and agree on the team's mission, purpose and objectives.
3. Innovative tactics that prevent your team from isolating itself from your organization's other units and departments.
4. Utilize 5 practical rules to assure team meetings are productive, motivating, and centered on results.
5. Discover the four keys to successful goal-setting, and learn how to incorporate them into your team structure.
6. Sound guidelines that pinpoint conflict within a team, and lead you to firm and quick solutions.
7. Guarantee team victory by including six key points in every well-rounded action plan – who, what, when, where, priority and resources.
8. How to be certain that every team member understands the group's goals – and what it means to their jobs on a daily basis.
9. How to consult with and include everyone you need to reach your objectives – no matter how your organization is structured.
10. Identify exactly which decisions should – and should not – require your team's consensus.
11. Sound direction to help you identify and harness each team member's unique strengths and assets.
12. Forthright advice on "favoritism" – how to recognize it, handle it and eliminate it from your team.
13. How to encourage organization-wide support for your team through positive, ongoing communications.
14. Firm steps to take when a team member challenges you in front of others.
15. Avoid damaging gripes and complaints with unique strategies that keep your team solution-oriented.
16. How to get individuals on your team to fully support group projects they disagree with – even innovative, risky undertakings.
17. How to celebrate successes so that all team members share equally in the victory.
18. How to incorporate a "code of ethics" that promotes team trust and harmony.
19. Learn essential motivational strategy that prevents business or organizational conditions you can't change from destroying your group's morale.
20. Use job interviewing tactics that pinpoint the best "team players" at the hiring stage.

How to Become a "Coach" and Build a Successful, Super-Productive, Winning Team.

This seminar equips you with the strategies and tactics that will immediately begin transforming your work group into a team that works together and wins in an inspiring atmosphere of energy and enthusiasm.

Invest just one day of your time, and the benefits of team-building will change your life as a manager; you'll witness an exciting synergy among your employees, mutual support and backup within your group, a sense of interdependence and exchange, and most of all – incredible productivity for your team as a unit.

This workshop concentrates on practical, realistic tips and strategies that guarantee you'll be equipped to begin building your winning team immediately. And, you'll learn how to deal with the day-to-day roadblocks managers face when building and managing teams.

This innovative, one-day seminar is 100% guaranteed – and is coming to your area soon. Enroll today!

Is It Really Possible to Accomplish All This? In Just One Day?

Yes – for one simple reason. We don't waste your time with impractical theories or irrelevant concepts. Our seminars concentrate on practical, how-to information that you can put to work right now, in your own workplace. That means tips, techniques, specific "how-to's" and strategies that will have an immediate impact on your team's effectiveness.

Set aside just one day, and you'll be able to "coach" your team to greater accomplishment, success and productivity. Your work group will buzz with enthusiasm and a team synergy that will amaze you. Why? Because training is knowledge, and knowledge is *power* – the power to meet and exceed the challenges that you and your team face everyday.

Comprehensive Program Agenda:

9:00 a.m. to 4:00 p.m.

A Powerful 9-point Action Plan for Building, Managing and Evaluating Your Team.

POINT 1: THE RIGHT INGREDIENTS FOR A WINNING TEAM

- Discover the secrets of well-known, successful teams that can be incorporated in any type of work group.
- The team versus the work group: pinpoint the specific characteristics that teams share and work groups lack.
- Understand how the team concept leads to success in today's business environment.
- Learn the 4 key benefits teamwork delivers that traditional work groups do not.

POINT 2: ASSESSING YOUR CURRENT TEAM EFFECTIVENESS

- Use a unique chart to assess your team on 10 key points and illuminate your present strengths and weaknesses.
- Uncover hidden strengths and weaknesses you may not be aware of with a step-by-step method for team self-assessment.
- How to identify the special, unique characteristics of your team that can contribute significantly to the group's synergy.
- Pinpoint existing team "building blocks" with a simple grid that precisely interprets your group's present methods of operation.

POINT 3: THE SUCCESSFUL TEAM'S WORKING MODEL

- Formulate a blueprint for team-building that you can use every day.
- Employ a 6-step checklist that incorporates every essential ingredient for smooth team operation.
- Compare the winning team model to your group's present operation and identify your team's "must-improve" areas.

POINT 4: DEFINING TEAM GOALS

- Use the "who, what, when, where, how?" action plan that turns team goals into results.
- 3 important questions you must answer before determining your team's goals.
- 4 key ingredients for setting realistic goals that everyone can understand.

- Discover the one-word question that will help every team member define goals more clearly.
- Employ a smart "fill-in-the-blank" formula that assures you'll set specific, reachable goals.

POINT 5: ESTABLISHING TEAM ROLES AND STRUCTURE

- Establish roles that emphasize job responsibilities and team contributions.
- Practical steps to help you master the 2 most essential qualities of an effective team leader.
- The manager's 4-step plan to lead a work group through the transition period to teamwork.
- Learn the 12 specific duties your team must assume – and how to guarantee they're effectively covered.
- Initiate 6 functional steps that get team results – from small tasks to major projects.
- How to get people to resolve personal differences and work out individual conflicts among themselves.

POINT 6: CLARIFY YOUR TEAM'S RULES AND RESPONSIBILITIES

- Learn which decisions you should make and which should be made by team consensus.
- How to handle aggressive types that intimidate and confront other team members.
- Clever remedies for "snipers" – team members that use "put-downs" to make other team members look bad.
- Firm actions to take when cliques and coalitions support opposing viewpoints.
- Create a "code of ethics" for your team that stops most conflict before it begins.

POINT 7: INTEGRATING INDIVIDUAL PERSONALITIES

- Evaluate and utilize the best personality traits of each individual for the good of the entire team.
- How to integrate team flexibility that minimizes the risk of "personality conflicts."
- Discover personality traits you may have that can frustrate members of your team.
- Learn to recognize the subtle signs of personality conflict and take corrective action early.

- Effective approaches that help you handle the most difficult people on your team.

POINT 8: MANAGING THE TEAM'S PERFORMANCE

- Learn the 8 crucial do's and don'ts for managing team performance.
- How to get support for your team from other employees, departments and support staff – and build an organization-wide respect for your team.
- 4 key points to make when communicating positive feedback – to one member or the entire group.
- Avoid hard feelings and negative attitudes by confronting trouble-makers the right way.
- Identify and eliminate policies that encourage unhealthy competition within a team.
- A step-by-step guide to meetings that motivate, inspire and prompt results from your entire team.
- Proven advice for team "brainstorming" sessions that produce more good ideas than you ever thought possible.
- How to produce an "action plan" that keeps every team member involved and informed.

POINT 9: EVALUATING YOUR TEAM'S EFFECTIVENESS

- 7 questions you must answer when evaluating a team's success.
- How to determine if your team is overemphasizing certain activities at the expense of others.
- Smart communication tips that ensure higher-ups recognize the accomplishments of your team.
- A powerful "self-evaluation" test that every member of your team can use to assess themselves.
- How to use team successes – and failures – as motivation for future performance.
- Utilize a simple chart to measure your team's performance against the "winning team model."

On-site Seminars

AN EFFECTIVE WAY TO GET THE FULL SUPPORT OF YOUR TEAM.

If you'd like to get a quick start and involve your entire group in the team-building process, investigate our on-site seminars.

Your entire team – and others within your organization – will benefit greatly from this program; group training is a highly effective way to assure a successful start for your team-building goals, with the full support and enthusiasm of your work group.

Our on-site programs bring the seminar to your organization, and customize it for your special needs – at a remarkably affordable price. Call 1-800-255-6278 for more information about on-site seminar training.

REGISTRATION INFORMATION

The fee for this seminar is $99 per person. (For five or more attending the same seminar, the fee per person is $89.) You may cancel your registration up to seven days before the seminar. Your registration fee will be refunded less a $10 enrollment charge. If you need to cancel less than seven days prior to the seminar, you may 1) send a substitute from your organization or, 2) transfer your registration to another seminar of your choice within 12 months.

The fee includes all workbook materials, seminar instruction and refreshment breaks. You may enjoy lunch on your own. Registration is permitted the day of the seminar on a space-available basis. Program hours are 9:00 a.m. to 4:00 p.m.

For fast registration call 1-800-255-6139 toll-free between 7:00 a.m. and 7:00 p.m. Monday through Friday, and Saturday 8:00 a.m. to 5:00 p.m. CST. To fax your registration, fax to (913) 384-2637. Our fax line is open 24 hours a day, seven days a week.

CONTINUING EDUCATION CREDITS

"How to Turn Your Work Group into a Winning Team" is approved for 0.6 CEU's of continuing education credit. Check with your local board for verification.

TAX DEDUCTION

All expenses of Continuing Management Education (including registration fees, travel, meals and lodging) taken to maintain and improve professional skills are tax deductible (Treas. Reg. 1.162-5 Coughlin vs. Commissioner, 203 F2d 307).

An

Innovative, One-day Seminar that Shows You How to Lead Your Workers to Greater Accomplishment, Success and Super-charged Productivity!

How To Turn Your Work Group Into A Winning Team

- Develop a work team that literally buzzes with enthusiasm, positive energy and a sense of purpose.
- Watch productivity and accomplishment soar to unheard-of levels – regardless of obstacles that now stand in your way.
- Become an intuitive group leader and earn the respect of each team member.
- Help every employee develop a "teamwork" attitude – even those who've been difficult to manage in the past.
- *And much more!* See inside for full details.

Register Today!
Call Toll-Free
1-800-255-6139

OR

Mail the registration form below

OR

FAX your registration, call
913-384-2637

LOCATIONS AND DATES

Birmingham, AL—Mar. 13
Seminar No. 80011
The Tutwiler Hotel
Park Place at 21st North

Atlanta, GA—Mar. 12
Seminar No. 80002
Georgia World Congress Ctr.
285 International Blvd.

Marietta, GA—Mar. 11
Seminar No. 80219
Marietta City Club
510 Powders Spring Street

Boston, MA—Mar. 5
Seminar No. 80013
Tremont House
275 Tremont Street

Brockton, MA—Mar. 7
Seminar No. 80176
Holiday Inn Randolph
1374 N. Main Street

Springfield, MA—Mar. 26
Seminar No. 80094
Holiday Inn Holyoke
245 Whiting Farms Road
Holyoke, MA

Woburn, MA—Mar. 6
Seminar No. 80246
Howard Johnson Lodge
I-93 & Montvale Avenue

Worcester, MA—Mar. 25
Seminar No. 80164
Sheraton Inn Worcester &
Conference Center
500 Lincoln Street

Oklahoma City, OK—Mar. 7
Seminar No. 80068
Howard Johnsons Hotel
5301 North Lincoln Blvd.

Providence, RI—Mar. 8
Seminar No. 80076
Holiday Inn Downtown
I-95 at Atwells

Austin, TX—Mar. 5
Seminar No. 80004
Holiday Inn Northwest Plaza
8901 Business Park Drive

Dallas, TX—Mar. 8
Seminar No. 80027
Cityplace Conference Center
2711 North Haskell Suite 100

Fort Worth, TX—Mar. 6
Seminar No. 80031
Holiday Inn South &
Conference Center
100 Alta Mesa East Blvd.

Houston, TX—Mar. 14
Seminar No. 80034
Marriott Hotel-Astrodome
2100 S. Braeswood & Greenbriar

San Antonio, TX—Mar. 4
Seminar No. 80087
Holiday Inn Riverwalk
217 North St. Mary's Street

REGISTRATION FORM – HOW TO TURN YOUR WORK GROUP INTO A WINNING TEAM

FRED PRYOR SEMINARS
A DIVISION OF PRYOR RESOURCES, INC.
P.O. Box 2951, Shawnee Mission, KS 66201

☐ **YES!** I'm ready to learn how to build a winning team – for only $99. Please enroll me today!

SEMINAR
Seminar City:
Seminar Date: Seminar No:

WHO WILL BE ATTENDING
Mr/Mrs./Ms: Title:
Mr/Mrs./Ms: Title:
Approving Mgr's. Name: Title:
Please list additional names on a separate sheet

YOUR ORGANIZATION
Organization:
Address:
City: St: Zip:
Tele: ☐ This is to confirm my phone registration.

FUTURE PROGRAMS
☐ I am interested in knowing about future programs. Please add me to your mailing list.

Attention: Mail Room Personnel (or Addressee) - Please Reroute if Necessary!

METHOD OF PAYMENT Please check one of the following:
1. ☐ Our Purchase Order is attached
   P.O. #
2. ☐ Bill my organization
   Attention
3. ☐ Registration fee enclosed
   Check #   Amount $

4. ☐ Charge to: ☐ Am. Express
   ☐ Mastercard
   ☐ Visa
Acct. no.
Exp. date
Signature

Important: Please send your payment now. Tuition is due before the seminar.
Please make checks payable to Pryor Resources, Inc. Please return this form to:
Pryor Resources, Inc., P.O. Box 2951, Shawnee Mission, KS 66201.
Please do not remove the mailing label.

T=891

Sri Lanka, Coast Conservation Department. 1988. **Coastal zone management plan.** Columbo, Sri Lanka: Coast Conservation Department, Ministry of Fisheries.

Steers, J. 1978. Saving the coast: The British experience. **Coastal Zone Management Journal** 4(1/2):7-23.

Stoddart, R. and Ferrari, J. 1983. Aldabra Atoll: A stunning success. **Nature and Resources** 19(1):20-28.

Suman, D. 1987. The management of coastal zone resources in Panama. In **Coastal Zone '87, Proceedings of the Fifth Symposium on Coastal and Ocean Management,** vol. 1, 1130-1145. New York: American Society of Civil Engineers.

Sumardja, E., and Wind, J. 1982. Nature conservation and rice production in Dumoga, North Sulawesi. Paper presented at the World National Parks Congress, October 11-12, 1982, Bali, Indonesia.

Susskind, L., and Cruickshank, J. 1987. **Breaking the impasse: Practical approaches to resolving public disputes.** New York: Basic Books.

Susskind, L. and McCreary, S. 1985. Techniques for resolving coastal resource management disputes through negotiation. **Journal of the American Planning Association** 51(3):365-374.

Susskind, L., and McMahon, J. 1985. The theory and practice of negotiated rulemaking. **Yale Journal on Regulation** 3(Fall):133-165.

Swanson, G. 1975. Coastal zone management from an administrative perspective: A case study of San Francisco Bay Conservation and Development Commission. **Coastal Zone Management Journal** 2(2): 81-102.

Sweden, Ministry of Physical Planning and Local Government. 1971. **National physical plan: Management of land and water.** Stockholm: Government Printing Office.

Taufik, A. 1987. Marine resource management in the Java Sea. In **Coastal Zone '87, Proceedings of the Fifth Symposium on Coastal and Ocean Management,** vol. 4, 4448. New York: American Society of Civil Engineers.

Templet, P. 1986. American Samoa: A coastal management area model for developing countries. **Coastal Zone Management Journal** 13(3/4):241-265.

Thailand, Office of Coastal Land Development. 1979. Establishment of a technological research center for coastal land development and management in Thailand. In **Proceedings of the Workshop on Coastal Area Development and Management in the Pacific,** edited by M. Valencia, 103-110. Honolulu: East-West Center and the University of Hawaii.

References and Bibliography

Thatcher, P., and Meith-Avcin, N. 1980. The oceans: Health and prognosis. In **Ocean Yearbook 1,** edited by E. Borgese and N. Ginsburg, 293-339. Chicago: University of Chicago Press.

Tolentino Jr., A. 1987. Philippine coastal zone management: Organizational linkages. In **Coastal Zone '87, Proceedings of the Fifth Symposium on Coastal and Ocean Management,** vol. 1, 699-713. New York: American Society of Civil Engineers.

Tortell, P. 1981. **New Zealand atlas of coastal resources.** Wellington: Government Printing Office.

Towle, E. 1985. The island microcosm. In **Coastal resources management: Development case studies,** 589-738. Coastal Management Publication No. 3, NPS/AID Series. Washington, D.C.: U.S. National Park Service and U.S. Agency for International Development

Travis, W. 1980. Coastal program evaluation from a state perspective. In **Coastal Zone '80, Proceedings of the Second Symposium on Coastal and Ocean Management,** vol. 1, 451-469. New York: American Society of Civil Engineers.

Trust Territory, Division of Lands. 1979. Coastal resources and environment: Trust Territory of the Pacific Islands. In **Proceedings of the Workshop on Coastal Area Development and Management in the Pacific,** edited by M. Valencia, 77-81. Honolulu: East-West Center and the University of Hawaii.

Turner, R. 1985. Coastal fisheries, agriculture and management in Indonesia: Case studies for the future. In **Coastal resources management: Development case studies,** 373-433. Coastal Management Publication No. 3, NPS/AID Series. Washington, D.C.: U.S. National Park Service and U.S. Agency for International Development.

UNEP. **See** United Nations Environment Program.

UNESCO. 1980. **Marine science and technology in Africa: Present state and future development.** Unesco Reports in Marine Science. Paris: UNESCO.

UNESCO and Intergovernmental Oceanographic Commission. 1979. Workshop on coastal area management in the Caribbean region, summary report. Workshop report, No. 26. Paris: UNESCO.

United Arab Emirates. 1976. **Coastal development planning study.**

United Nations. 1978. Mediterranean Action Plan and Final Act. In **Conference of the Plenipotentiaries of Coastal States of the Mediterranean Region for the Protection of the Mediterranean Sea.** New York: United Nations.

United Nations. 1980. **Conference of the Plenipotentiaries of Coastal States of the Mediterranean Region for the Protection of the Mediterranean Sea Against Pollution from Land Based Sources.** New York: United Nations.

United Nations. 1982. **Conference of the Plenipotentiaries on the Protocol Concerning Mediterranean Specially Protected Areas.** New York: United Nations.

United Nations. 1983. Final Act. In **Conference of the Plenipotentiaries on the Protection and Development of the Marine Environment of the Wider Caribbean region.** New York: United Nations.

United Nations, Department of Technical Cooperation for Development. 1983. **Experiences in the development and management of international river and lake basins.** Proceedings of the U.N. Interregional Meeting of International River Organizations, Dakar, Senegal. New York: United Nations.

United Nations Development Programme and the Government of Yugoslavia. 1968. Physical development for the south Adriatic coast. Split, Yugoslavia: Shankland, Cox & Associates and the Urbanisticki Zavod Dalmacije.

United Nations, Economic and Social Council, Committee for Programme and Coordination. 1983. **Cross-organizational programme analysis of the activities of the United Nations system in marine affairs.** Report of the Secretary General. New York: United Nations.

United Nations Environment Program. 1980. **Declaration of environmental policies and procedures relating to economic development.** Geneva: UNEP.

United Nations Environment Program. 1982a. **Marine and coastal area development in the East African region.** UNEP Regional Seas Reports and Studies No. 6. Geneva: UNEP.

United Nations Environment Program. 1982b. **Achievements and planned development of UNEP's Regional Seas Programme and comparable programmes sponsored by other bodies.** UNEP Regional Seas Reports and Studies No. 1. Geneva: UNEP.

United Nations Environment Program. 1982c. **Environmental problems of the East African Region: An overview.** UNEP Regional Seas Reports and Studies No. 12. Geneva: UNEP.

United Nations Environment Program. 1983. **Action plan for the Caribbean Environment Programme.** UNEP Regional Seas Reports and Studies No. 26 (June). Geneva: UNEP.

United Nations Environment Program. 1985. **The Siren: News from UNEPs Regional Seas Program,** No. 3.

United Nations, Joint Group of Experts on the Scientific Aspects of Marine Pollution (GESAMP). 1980. **Marine pollution implications of coastal area development.** Reports and Studies, No. 11. New York: United Nations.

United Nations, Ocean Economics and Technology Branch, Department of International Economic and Social Affairs. 1979. **Marine and coastal area development in the wider Caribbean area: Overview study.** Doc# E/CEPAL/PROY 3/L. New York: UNOETB.

United Nations, Ocean Economics and Technology Branch, Department of International Economic and Social Affairs. 1982a. **Coastal area management and development.** Oxford: Pergamon Press.

United Nations, Ocean Economics and Technology Branch, Department of International Economic and Social Affairs. 1982b. **Technologies for erosion control.** Doc# SC/EFA/116. New York: UNOETB.

United Nations, Ocean Economics and Technology Branch, Department of International Economic and Social Affairs. 1985. Materials collected by the Office of Ocean Affairs and Law of the Sea.

United Nations, Office of Ocean Affairs and Law of the Sea. Forthcoming. **Coastal and marine multiple use conflicts, problems and approaches** (title tentative). New York: United Nations.

U.S. Agency for International Development. 1979. **Environmental and natural resource management in developing countries: A report to Congress,** vol. 1. Washington, D.C.: U.S. Department of State.

U.S. Agency for International Development. 1981. **Report on the Coastal Zone Management Seminar-Workshop.** November 5-6, 1981, Manila. USAID Mission, Philippines.

U.S. Agency for International Development. 1982. **Potential strategies and target countries for coastal resources management activities.** Washington, D.C.: USAID/S&T/FNR.

U.S. Agency for International Development and U.S. Department of Commerce, National Oceanic and Atmospheric Administration. 1987. **Caribbean marine resources opportunities for economic development and management.** Washington, D.C.: U.S. Department of Commerce.

U.S. Commission on Marine Science, Engineering and Resources. 1969. **Our nation and the sea.** Washington, D.C.: U.S. Government Printing Office.

U.S. Department of Commerce, Office of the Inspector General. 1983. **Opportunities to improve federal oversight responsibilities of states' coastal zone management programs to assure more effective results.** Washington, D.C.: U.S. Government Printing Office.

U.S. Department of Interior. 1970. **The national estuarine pollution study.** U.S. Senate, 91st Congress, 2nd Session, Document 91-58. Washington, D.C.: U.S. Government Printing Office.

U.S. Department of State, Bureau of Intelligence and Research. 1983. **Status of the world's nations.** Washington, D.C.: U.S. Government Printing Office.

U.S. Environmental Protection Agency, Office of International Activities. 1977. **International environmental issues: A preliminary resource guide.** Washington, D.C.: Office of International Activities.

U.S. National Academy of Sciences, Board on Science and Technology and the Ocean Policy Committee. 1982. Working papers prepared for the International Workshop on Coastal Resource Management, February 8-10, La Jolla, California.

U.S. National Oceanic and Atmospheric Administration, Department of Commerce, National Oceanic and Atmospheric Administration. 1981. **The Federal coastal programs review: A report to the president.** Washington, D.C.: National Oceanic and Atmospheric Administration.

U.S. National Oceanic and Atmospheric Administration, Department of Commerce, Office of Coastal Zone Management and California Coastal Commission. 1981. **Final environmental impact statement: Tijuana River National Estuarine Sanctuary.** Washington, D.C.: U.S. Government Printing Office.

U.S. National Oceanic and Atmospheric Administration, Department of Commerce, Office of Coastal Zone Management. 1982. **Biennial report to the Congress on coastal zone management: Fiscal years 1980 and 1981.** Washington, D.C.: U.S. Government Printing Office.

University of Rhode Island and U.S. Agency for International Development. 1987. **Coastal Resources Management Project: Prospectus.** Kingston, RI: University of Rhode Island Coastal Resources Center.

UNOETB. See United Nation, Ocean Economics and Technology Branch, Department of International Economic and Social Affairs.

Uruguay, Direccion Nacional di Relaciones Publicas. 1983. Government of Uruguay. Brochure.

USAID. See U.S. Agency for International Development.

Valencia, M., ed. 1979. **Proceedings of the Workshop on Coastal Area Development and Management in Asia and the Pacific, Manilla.** Honolulu: East-West Center and the University of Hawaii.

Vallejo, S. 1983. El marco institucional para el desarrollo y ordinacion de los recursos costeros y marinos: Problemas y experiencias. Working paper for a marine resources planning seminar, May, 1983, Mexico City.

Vallejo, S. 1987. Report: Seminar on the integrated development and management of coastal areas: A pilot experience. **Coastal Management Journal** 15(1):89-97.

Vallejo, S. 1989. Development and management of coastal and marine areas: An international perspective. **Ocean Yearbook 7, edited by E. Borgese, N. Ginsburg, and J. Morgan**, 205-222. Chicago: University of Chicago Press.

Vallejo, S., and Caparro, L. 1981. **Preliminary information on the resources, uses and problems of the Ecuadorian coastal area.** New York: United Nations Department of International Economic and Social Affairs, Ocean Economics and Technology Branch.

Victoria, Ministry for Planning and the Environment. 1988. **A coastal policy for Victoria.** Victoria, Australia.

Vicuna, F., ed. 1986. **Preservacion del medio ambiente marino.** Santiago: Instituto de Estudios Internacionales de la Universidad de Chile.

Waite, C. 1980. Coastal management in England and Wales. In **Comparative marine policy: Perspectives from Europe, Scandinavia, Canada, and the United States**, 57-64. New York: Praeger Special Studies.

Warren, R., Weschler, L., and Rosentraub. 1977. Local-regional interaction in the development of coastal land use policies: A case study of metropolitan Los Angeles. **Coastal Zone Management Journal** 3(4):331-362.

Wetterberg, G. 1982. The exchange of wildland technology: A management agency perspective. Paper presented at the World National Parks Congress, October 11-12, 1982, Bali, Indonesia.

White, A., and Savini, G. 1987. Community based marine reserves: A Philippine first. In **Coastal Zone '87, Proceedings of the Fifth Symposium on Coastal and Ocean Management**, vol. 1, 2022-2038. New York: American Society of Civil Engineers.

Wiggerts, H., and Koekebakker, P. 1982. Coastal planning and management in the Netherlands. **Ekistics** 293:143-149.

Wilcox, E. 1987. Marine park planning in the Third World: Haiti case study. In **Coastal Zone '87, Proceedings of the Fifth Symposium on Coastal and Ocean Management**, vol. 1, 3568-3579. New York: American Society of Civil Engineers.

World Bank. 1982. **Environmental requirements of the World Bank.** Washington, D.C.: Office of Environmental Affairs, Projects Advisory Staff.

World Environment Report. 1981a. October 30.

World Environment Report. 1981b. November 30.

World Environment Report. 1982a. June 30.

World Environment Report. 1982b. December 30.

World Environment Report. 1983a. May 15.

World Environment Report. 1983b. June 15.

World Environment Report. 1983c. June 30.

World Environment Report. 1983d. October 30.

Zamora, P. 1979. The coastal zone management program of the Philippines. In **Proceedings of the Workshop on Coastal Area Development and Management in the Pacific,** edited by M. Valencia, 85-88. Honolulu: East-West Center and the University of Hawaii.

Zhung, G. 1985. On the implementation of coastal law. In **Coastal Zone '85: Proceedings of the Fourth Symposium on Coastal and Ocean Management,** vol. 2, 2320-2334. New York: American Society of Civil Engineers.

Zile, Z. 1974. A legislative-political history of the Coast, Zone Management Act of 1972. **Coastal Zone Management Journal** 1(3): 235-274.

AUTHOR BIOGRAPHIES

Jens C. Sorensen is Principal, Jens Sorensen and Associates, and Adjunct Associate Professor in Marine Affairs at the University of Rhode Island. His Ph.D. is in environmental planning from the University of California at Berkeley. He has specialized in coastal management since 1968 and has carried out numerous studies in the United States (particularly California), Australia, Mexico, Costa Rica and South America and served as consultant to the United Nations, the Organization of American States, the U.S. Agency for International Development, and the U.S. National Oceanic and Atmospheric Administration.

Dr. Sorensen is the author of 36 articles, papers and reports on coastal management and impact assessment as practiced both in the United States and abroad. He has taught courses on ocean and coastal policy, regional environmental planning and impact assessment at three campuses in the University of California system, the Massachusetts Institute of Technology, and the University of Rhode Island.

During 1988 and 1989, with support from a Fulbright Scholarship, he has been assessing programs in Latin America for the sustainable development of coastal resources and environments.

Scott T. McCreary completed his Ph.D. at the Massachusetts Institute of Technology in the Department of Urban Studies and Planning and served as an Associate of the M.I.T.-Harvard Public Disputes Program. Presently he is a Lecturer in Environmental Planning in the Department of Landscape Architecture at the University of California at Berkeley. He has 12 years of experience in coastal planning and marine resource management. His international work includes co-authorship of "Prospects for Integrated Coastal Resources Management in West Africa" (USAID, 1988); research on transboundary environmental problems; impact assessment for the Caribbean; and training in environmental dispute resolution in the Caribbean, Canada, and Australia and New Zealand. As consultant to the New York Academy of Sciences, he mediated a complex negotiation to improve management of PCBs in the New York Harbor Region.

Dr. McCreary earned his Master's degree in Landscape Architecture and Environmental Planning at the University of California, Berkeley. With California's State Coastal Conservancy, he led the agency's work with non-profit organizations to restore wetlands, streams, and watersheds, and he launched the designation of the Tijuana River National Estuarine Sanctuary. He has taught university courses in coastal management, land use planning, environmental planning in developing countries, and dispute resolution and has authored over 30 articles and papers on these and other environmental policy topics.